AF334237

Introduction to Topological Semigroups

Anthony Connors Shershin

A Florida International University Book
UNIVERSITY PRESSES OF FLORIDA
Miami 1979

University Presses of Florida is the central agency for scholarly publishing of the State of Florida's university system. Its offices are located at 15 NW 15th Street, Gainesville, FL 32603. Works published by University Presses of Florida are evaluated and selected for publication by a faculty editorial committee of any one of Florida's nine public universities: Florida A&M University (Tallahassee), Florida Atlantic University (Boca Raton), Florida International University (Miami), Florida State University (Tallahassee), University of Central Florida (Orlando), University of Florida (Gainesville), University of North Florida (Jacksonville), University South Florida (Tampa), University of West Florida (Pensacola).

Published through the Imprint Series,
Monograph Publishing.
Produced and distributed by
University Microfilms International
Ann Arbor, Michigan 48106

Library of Congress Cataloging in Publication Data

Shershin, Anthony Connors.
   Introduction to topological semigroups.

   (Monograph publishing : Imprint series)
   Bibliography: p.
   Includes index.
   1. Topological semigroups. I. Title.

QA387.S46   512'.55   79-23432
ISBN 0-8130-0664-3

Dedication:

To my family: wife Carmen
and daughters Roxana, Alexandra and Tania

TABLE OF CONTENTS

PREFACE

The main purpose of this monograph is to present in a
simplified manner the basic results and methods employed in
topological semigroups. The startling algebraic conclusions
using topological hypotheses and the elegance of the proofs
are justification enough for a thorough study of this area
of mathematics. It is true that no attempt is made to include
all the basic results and methods of this field in this book
for to do so would necessitate a far more rigorous background
on the part of the reader than is the intent of this author.
I hope that this book will be judged on its clarity rather
than its completeness. For a more thorough treatment of
this area one may seek Mostert and Hoffmann's book [1] on
the subject.

A secondary objective is to specifically indicate how
two major areas of mathematics, algebra and topology, can
interact. In our topic, a set of objects is given simul-
taneously an algebraic structure (that of a semigroup) and
a topological structure. The topology affects the positioning
of the objects and the algebra states how the objects can
combine. To indicate how the positioning influences this
combining of objects is the object of this monograph. It is
in this regard that topological semigroups are characteristic
of what has become known as topological algebra. (Algebraic
topology, a related area, places its emphasis on the reverse

situation, namely, how positioning is influenced by the
combining of objects, that is, what topological results may
be obtained from algebraic structure.)

The algebraic notions are introduced as the theory
warrants because they are few in number. However, the
preliminary topological results are introduced in Chapter
1 because I feel that to require a thorough mastery of
topology is not essential before embarking upon a study in
this field. With this in mind I have isolated the necessary
basic topological concepts and results. It is the concepts
and results which will be needed and not the topological
methodology and so the proofs in Chapter 1 may be omitted
at the reader's discretion. Some of the lesser known and
more necessary topological results are given, together with
their proofs, as the theory develops in succeeding chapters.

This book is aimed at college undergraduates, graduates
with little topological background and teachers wishing an
introduction to modern techniques. Such a reader is expected
to have an elementary knowledge of set theory, functions and
relations, real numbers, high school algebra, and the logic
necessary to follow a simple mathematical proof. The core
of the text is Chapters 2, 3 and 4. The extent which Chapter
1 is included will depend on the background of the reader:
those persons very familiar with topology may skip Chapter 1;
those with limited knowledge of topology or those whose
memories time has made misty should, at the very least,

refresh their memories with a cursory reading; those with no knowledge of topology should read the first chapter carefully-- it is self-contained.

At this point it should be mentioned that the important topological concept of a metric space (all the more important here because every metric space is Hausdorff, the type of underlying space we restrict ourselves to in this text) has not been treated in this book, the reason being that it is not needed for the semigroup topics we discuss here. In fact, continuity of ordinary multiplication and addition has been relegated to an appendix because the proofs are metric in nature, that is, the classical ε, δ techniques are used. That is not to say that the idea of a metric is insignificant in the field of topological semigroups; rather, its usage is not elementary in nature. Consequently, the reader whose topo- logical experience is beginning here is encouraged in the future to read other texts for information concerning metric spaces.

Chapters 2-4 detail the fundamental structure results, both algebraic and topological. The dominant topological notions are compactness and connectedness and, algebraically, groups and ideals appear frequently. Chapter 5 begins to bridge the gap between the elementary nature of semigroups and a more sophisticated approach. The notion of a homo- morphism, fundamental to modern mathematics, is discussed in regards to its semigroup consequences and its usage in forming

new semigroups. The Rees quotient semigroup is introduced
at the end. Another possible candidate for a quotient
semigroup, namely, the quotient space $S/\mathcal{H}$ is examined in
Chapter 6.

The final Chapter, 7, differs from the first six in its
approach. Its primary goal is to stimulate the student to
further study of semigroups by touching upon a number of
significant, nonelementary topics. No mastery of the material
is expected and not all statements are proved. Primarily, it
is devoted to the construction of "new" semigroups from
previously known semigroups. Most significant is the initial
section on semigroup products which deals with cartesian
product spaces. Also, emphasis is placed upon verifying the
continuity of the semigroup multiplications which are introduced.

The reader who has a background in abstract algebra and
real variables is encouraged, after finishing this book, to
read more advanced works on topological semigroups in order
to achieve a complete view of this subject. In particular,
one should then view descriptions of monothetic and cylindrical
semigroups, one-parameter semigroups, and irreducible semigroups.

A word about the exercises at the end of Chapters 1-6.
They too are not an integral part of this book and they have
been included solely to stimulate reflection. Many are quite
easy and none require lengthy analysis. Some readers, in

particular, graduate students, may choose to omit a number
of the exercises, especially the simpler proofs; however, it
is suggested that everyone attempt all those exercises which
deal with examples, particularly those which ask the reader
to find examples. Exercises are purposely omitted from
Chapter 7 to emphasize the nonrigorous, expository presentation
of the material.

In order to indicate which results are not dependent upon
topology, I have inserted the term algebraic, in parentheses,
before each proposition whose proof is entirely algebraic.

This text can be adequately covered in one quarter if
the students have even a minimal topological background; two
quarters or one semester would be necessary for a class with
no topology experience.

I wish to take this opportunity to give credit to the
numerous people who have helped, in various ways, to make
this book possible. First and foremost among my creditors
is my graduate mentor, Dr. Alexander Doniphan Wallace, both
for his brilliant unpublished lecture notes entitled, "Project
Mob," (an early synonym for semigroup) and for his inspirational
guidance during my graduate studies. For these reasons all
that is correct and clearly presented in this book may be traced
to him as a source. (I fully accept all responsibility for
those portions which are incorrect or unclear.) Dr. Jane
Day's expository article [5] also proved of textual interest,
particularly in Chapters 6 and 7, as did a number of lectures

of hers that I have been privileged to attend. Special thanks go to my wife Carmen who, in addition to displaying patience and offering encouragement in my endeavor, read the entire manuscript and made significant comments. Thanks also go to my brother John Shershin; my colleagues, Professors Thomas Borrego, Frank Cleaver, Malcolm Gotterer, John Leeson, You-Feng Lin, Donald Rose and Leonard Soniat; students David Rose and Roger Taylor; and secretaries Mrs. Janice Auton and Mrs. Jeanne Disney; all these people have contributed, in one way or another, to the writing of this text. I am also grateful to Paul Mostert for his review of my original manuscript which I have revised to incorporate some of his observations.

Lastly, I wish to express my thanks to the Florida International University Research and Publications Committee, and specifically to two successive chairpersons of that committee, Anthony Maingot and Margaret Waid, for selecting this manuscript for publication and also to the Florida International Foundation and College of Arts and Sciences whose finances supported the preparation of the text in its present "camera-ready" form, suitable for immediate printing.

CHAPTER I. TOPOLOGICAL TOPICS

Section 1. Historical Introduction

Point-set Topology had its initial origins with Georg Cantor in 1870. At that time the studies dealt primarily with Euclidean spaces, i.e., ordinary 3-dimensional space and generalizations. The more abstract methods that we use today had their beginning impetus about 1914 when a number of mathematicians, notably Maurice Fechet and Felix Hausdorff, began to consider spaces without a distance function. In this chapter we too will not deal with metric spaces, i.e., spaces with a distance function, even though such spaces are fundamental to topology. There are a number of good elementary texts which treat the notion of a metric space well, [6] among others, and because of this and the fact that we will not need it in the following chapters, we omit this important concept.

Section 2. Set Theoretic Remarks

Before turning to the basic topological concepts, we will briefly mention an important set-theoretic axiom and three equivalent formulations.

If $\mathscr{C}$ is a collection of sets, then the set of all elements belonging to at least one set in $\mathscr{C}$, written $\{x; x$ in C for at least one C in $\mathscr{C}\}$, is called the <u>union</u> of the collection $\mathscr{C}$ and denoted by $\cup\mathscr{C}$ or $\cup\{C; C \in \mathscr{C}\}$.

Axiom of Choice. If for each element a in some set A,

X_a is a non-empty set, then there is a mapping f of A into
$U\{X_a$; a in A} such that f(a) is in X_a for each a in A. In
other words, this assumption allows us always to consider a
collection $\mathcal{C}$ of non-empty sets and to select one element from
each set C in $\mathcal{C}$. It may be mentioned that, although this
axiom is used by most mathematicians, nevertheless it is not
accepted by those known as the "intuitionist school".

There are a number of equivalent, though very different
formulations of this axiom. We will state, but not prove,
three of these forms. The first, Zorn's Lemma, will be of
significant use to us in our later work with semigroups;
the second, the Hausdorff Maximal Principle, will be needed
in the proof of Tychonoff's Product Theorem later in this
chapter; and the third, Well-Ordering Principle, is of
general mathematical interest. For additional formulations
of the Axiom of Choice and further discussion the reader may
refer to [3].

Before stating these alternative forms, a number of
preliminary definitions must be listed. During this enumeration, as is the common custom, the biconditional phrase "if
and only if" will be abbreviated to iff. Also, a in A will
sometimes be written a ε A.

A relation R on a set S is <u>antisymmetric</u> iff for all
distinct pairs a, b ε S, i.e., a $\neq$ b, it happens that aRb
implies b$\not R$a. A <u>partial</u> <u>order</u> is a reflexive, antisymmetric
and transitive relation. Two common partial orders are $\leq$

and set inclusion, $\subseteq$. If $\subseteq$ partially orders a family $\mathcal{F}$ of sets, M is a <u>maximal</u> member of $\mathcal{F}$ iff no member of $\mathcal{F}$ contains M as a proper subset. A <u>simple order</u> is a partial order in which each pair of elements is comparable, that is, aRb or bRa for all a, b ε S. A <u>tower</u> is a family $\mathcal{T}$ of subsets such that $\mathcal{T}$ is simply ordered by inclusion. U is an <u>upper bound</u> of a tower $\mathcal{T}$ iff T $\subseteq$ U for all T ε $\mathcal{T}$. In contrast to a maximal member, <u>U need not be in</u> $\mathcal{T}$. A <u>well ordered</u> set S is a simply ordered set in which every nonempty subset of S has a "smallest" element.

Zorn's Lemma. If each tower in a partially ordered set S has an upper bound, then there is a maximal element of S.

For those acquainted with linear algebra, it is noted that Zorn's Lemma supplies a simple proof of the fact that every nontrivial vector space has a basis.

Hausdorff Maximal Principle (HMP). If $\mathcal{F}$ is a family of sets and $\mathcal{T}$ is a tower in $\mathcal{F}$, then there is a maximal tower $\mathcal{M}$ "containing " $\mathcal{T}$ in $\mathcal{F}$.

Actually, there are two relations involved in the HMP as indicated by the quotation marks. The underlying relation is set inclusion in $\mathcal{F}$ and the secondary relation R is defined on the collection of towers in $\mathcal{F}$, namely, $\mathcal{T}$ R $\mathcal{T}'$ iff T ε $\mathcal{T}'$ for all T ε $\mathcal{T}$. $\mathcal{M}$ is maximal with respect to the relation R.

Well Ordering Principle. Every set can be well-ordered.

As a result of the Well Ordering Principle the set of

real numbers can be well ordered. It is easy to see that
the usual simple order $\leq$ will not suffice. Indeed, no one
has yet produced a well ordering for this set.

We wind up this section with miscellaneous set notation
and information. The set with no elements is termed the
empty set and denoted by ϕ. If A is a subset of a set S,
then the set of elements in S but not in A is termed the
complement of A and written as S\A.

If $\mathscr{C}$ is a collection of subsets of a set S, the set
{x; x ε C for every C ε $\mathscr{C}$} is called the intersection of
$\mathscr{C}$ and denoted by $\cap\, \mathscr{C}$ or $\cap${C; C ε $\mathscr{C}$}. One may verify that
S\ ($\cup\mathscr{C}$) = $\cap${S\C; C $\varepsilon\mathscr{C}$} and S\ ($\cap\mathscr{C}$) = $\cup${S\C; C ε $\mathscr{C}$}.
These results are known as De Morgan's Laws.

If f is a function, it is easy to verify that f(A $\cup$ B) =
f(A) $\cup$ f(B) and, more generally, $f(\underset{\alpha\varepsilon\Lambda}{\cup} A_\alpha) = \underset{\alpha\varepsilon\Lambda}{\cup} f(A_\alpha)$ for
any index set Λ. Also, f(A $\cap$ B) $\subseteq$ f(A) $\cap$ f(B) and equality
holds if and only if f is a one-to-one function, that is,
f(a) = f(b) only if a = b.

Section 3. Preliminaries

Let X be a set. A collection $\mathscr{T}$ of subsets of X is a
topology for X if and only if (i) the union of each sub-
collection of $\mathscr{T}$ is a member of $\mathscr{T}$ and (ii) the intersection
of each finite subcollection of $\mathscr{T}$ is a member of $\mathscr{T}$. The
pair (X, $\mathscr{T}$) is then called a topological space but, when the
topology is well understood, it is common to refer to "the
topological space X" or, "the space X". In addition, the

elements of a space X are called points. The members of $\mathcal{T}$ are called <u>open sets</u>. A subset C of X is <u>closed</u> iff X\C is open. Lastly, the <u>closure</u> A* of a set A is the smallest closed set containing A. It is noted that A is closed if and only if A = A*.

For illustrations we first list two trivial, but interesting, topologies: Let X be any set and $\mathcal{D}$ be the family of all subsets. (X,$\mathcal{D}$) turns out to be a topological space and $\mathcal{D}$ is referred to as the <u>discrete topology</u>. Now let X again be any set but this time consider $\mathcal{N} = \{\phi,X\}$, that is, $\mathcal{N}$ consists solely of the nonproper subsets of x. $\mathcal{N}$ is the <u>indiscrete topology</u> for X.

We now turn our attention to the most interesting and most frequently studied set in mathematics, the set R of real numbers. Let U be a subset of R such that for each $r \in U$ there correspond real numbers a, b with the properties that $a < r < b$ and the interval $(a, b) \subseteq U$. The collection $\mathcal{U}$ of all sets U of this type form a topology for R known as the <u>usual topology for R</u>. When we speak of the space R we are referring to this usual topology unless some other topology is explicitly named.

A set X with the discrete topology and the space R with the usual topology lead us to a general type of space in which every two distinct points may be separated by two members of the topology. Such a space is said to be Hausdorff, in honor of Felix Hausdorff, or more anonymously, a T_2-space. More formally, $(X,\mathcal{T})$ is a <u>Hausdorff space</u> (or

T_2-space) iff for every pair of distinct points p and q

(i.e., $p \neq q$) there exist two sets U and V in $\mathcal{T}$ such that

$p \in U$, $q \in V$ and $U \cap V = \phi$. The term T_2 is related to other

types of "separation" of points. Since we will not discuss

other types of separation, the term Hausdorff will be used

exclusively throughout the rest of this book for the sake

of consistency and to avoid possible confusion for those

readers not too familiar with topology.

In Hausdorff spaces all finite subsets are closed.

It has been said, more than once, that the only "useful"

spaces are Hausdorff. One reason for such a comment is that

all metric spaces are Hausdorff. All topological semigroups

studied in this book will be Hausdorff.

Let $A \subset X$ and $(X,\mathcal{T})$ be a topological space. The family

$\mathcal{F} = \{A \cap T;\ T \in \mathcal{T}\}$ constitutes a topology for A known as

the <u>relative topology</u>. $(A,\mathcal{F})$ is called a <u>subspace</u> of $(X,\mathcal{T})$.

When a subset of a space is said to have a specific

topological property it is implicit that the subset is being

considered as a subspace. For example, one of the exercises

at the end of this chapter reads, a subset of a Hausdorff

space is Hausdorff. It could be more precisely written as,

a subset S of a Hausdorff space, S being considered as a

subspace with the relative topology, is Hausdorff.

However, there is no ambiguity when properly

understood and this convention has the important advantage of brevity. This same idea will occur later when we speak of a "compact subset" and a "connected subset". Actually, it is possible to discuss topological properties of subsets of spaces without the notion of a subspace [6] and at first this may be helpful; however, such an approach may cause confusion as a student progresses to higher levels of sophistication in topology.

Section 4. Continuous functions.

One cannot consider mathematical systems for very long without discussing relations, in general, and continuous functions, in particular. We have reached that juncture in topology:

If f is a function from a space $(X, \mathcal{T})$ to a space $(Y, \mathcal{V})$, then for any subset $B \subseteq Y$ we define $f^{-1}(B) = \{x \in X; f(x) \in B\}$. With the same notation, f is <u>continuous</u> iff $f^{-1}(V)$ is open for each $V \in \mathcal{V}$, that is, verbally, the inverse image of each open set is open. We shall prove two important equivalent formulations of continuity:

1.1. <u>Proposition</u>. Let f be a function from a space X into a space Y. f is continuous if and only if

(a) the inverse image of each closed subset of Y is closed in X; or

(b) for each subset A of X, $f(A^*) \subseteq [f(A)]^*$.

Proof. We will prove that (a) is equivalent to continu-

8

ity and (b) is equivalent to (a). Suppose f is continuous and let B be a closed set of Y. Then $f^{-1}(Y\backslash B)$ is open and it is easy to verify that $f^{-1}(Y\backslash B) = X\backslash f^{-1}(B)$. Thus $f^{-1}(B)$ is closed. Conversely, suppose that (a) holds. Let V be open in Y. Then $f^{-1}(Y\backslash V) = X\backslash f^{-1}(V)$ is closed and so $f^{-1}(V)$ is open.

By convention $[f(A)]*$ is usually abbreviated to $f(A)*$.

Suppose (a) holds and let $A \subseteq X$. $f^{-1}(f(A)*)$ is closed by assumption. As a result, since $A \subseteq f^{-1}f(A) \subseteq f^{-1}(f(A)*)$ and using the definition of closure, we have $A* \subseteq f^{-1}(f(A)*)$. Therefore, $f(A*) \subseteq ff^{-1}(f(A)*) \subseteq f(A)*$. Conversely, suppose (b) holds, consider any closed set B of Y and let $D = f^{-1}(B)$. $f(D) \subseteq B$ and so $f(D)* \subseteq B* = B$. Then, by hypothesis, $f(D*) \subseteq B$ and, consequently, $D* \subseteq f^{-1}f(D*) \subseteq f^{-1}(B) = D$. Thus $f^{-1}(B)$ is closed and the proof is completed.$\square$

A continuous function is often called a <u>mapping</u> or a <u>map</u>.

If a space X has the discrete topology, the largest possible topology, then every function with domain X is continuous. But if we want the smallest topology $\mathscr{S}$ for a set X so that a particular function f, with range $(Y, \mathscr{V})$, is continuous, then it is sufficient to take $\mathscr{S} = \{f^{-1}(V); \ V \in \mathscr{V}\}$. The emphasis in this paragraph so far has been on the domain space X. Let us reverse the situation and ask if, given a function g, may we always topologize Y in such a manner that g is continuous? The answer is yes, namely, $\mathscr{V} = \{V; \ g^{-1}(V)$

is open in X}. This is the quotient topology for Y and it is the largest topology for Y for which g is continuous.

Let X, Y and Z be spaces and f:X → Y and g:Y → Z. We define the composition function gf:X → Z by [gf](x) = g(f(x)).

1.2. Proposition. If f:X → Y and g:Y → Z are both continuous, then the composition function gf is continuous.

Proof. Let U be an open set of Z. It is easily verified that $(gf)^{-1}(U) = f^{-1}(g^{-1}(U))$ and so the result follows easily.□

We end this assortment of continuity topics by mentioning how topological spaces are "alike": From an abstract viewpoint, two spaces X and Y are homeomorphic, i.e., topologically alike, iff there exists a continuous, 1-1 function h:X → Y with the property that the inverse function h':Y → X defined by h'(y) = x iff h(x) = y, is also continuous. Such a function h is a homeomorphism. Homeomorphisms induce an equivalence relation among topological spaces. For a visual, intuitive interpretation of topological spaces see [1], [5] and [7].

One definition of topology, as a mathematical discipline, is the study of those properties of point sets which are invariant, i.e., unchanged, by homeomorphisms. Later we will discuss two important such properties, compactness and connectedness. However, familiar properties such as size, length, area and volume are not invariant under homeomorphisms.

As examples, we mention that any open interval, such as (0,1), is homeomorphic to the real line R and the closed interval [0,1] is topologically equivalent, i.e., homeomorphic, to any other closed interval [a,b] regardless of its length. In the plane R^2, an open disk, for instance, one (with the origin as center) of the form $x^2 + y^2 < a^2$, which is an open set, is homeomorphic to any open square, i.e., one not including the boundary; moreover, any two such open disks, or squares, are topologically equivalent. In fact, an open disk is homeomorphic to the plane itself! Also, a circle, say $x^2 + y^2 = a^2$ for some constant a, is homeomorphic to any ellipse. Analogous results hold in 3-dimensions. Sets in one dimension can even be homeomorphic to sets in another dimension. For instance, a circle, in the plane, with one point deleted is homeomorphic to an open interval of real numbers. Also, by means of a useful technique known as the stereographic projection, one can show that, with a single point taken out, the surface of a (3-dimensional) ball is topologically equivalent to the entire plane R^2.

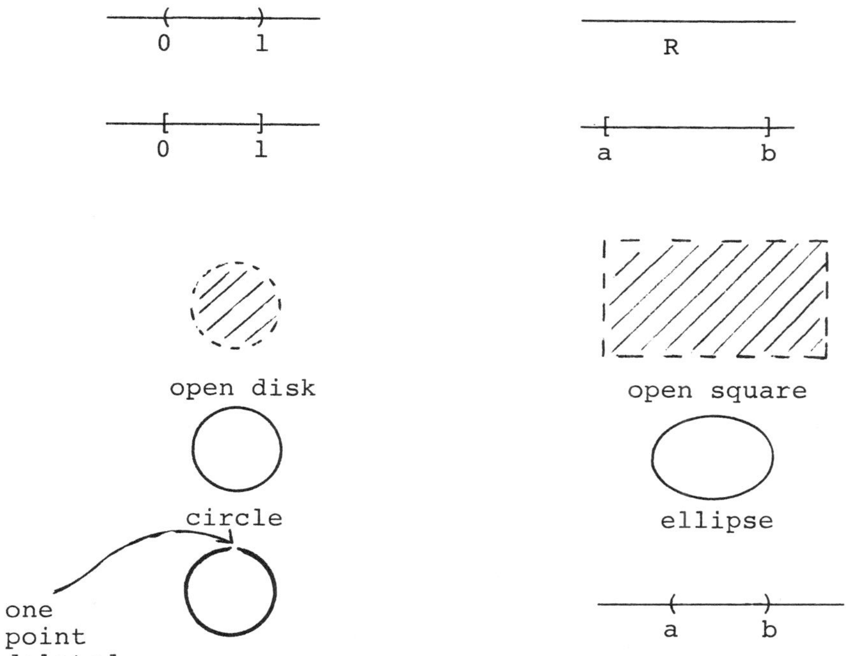

Figure 1.1 Pairs of Homeomorphic Sets

Intuitively, two geometric sets are topological "look-alikes", i.e., are homeomorphic, if one set can be changed into the other by stretching some of its parts, shrinking others or even bending or twisting (but never tearing). In this connection, topology has been referred to as the rubber geometry because of the pliable nature of rubber.

On the other hand, we find that the real line is not homeomorphic to the plane and neither of these is (topologically) equivalent to 3-dimensional space R^3. In addition, the real line is not homeomorphic to [0,1] and neither of these is homeomorphic to any disconnected subset of R, such as {0,1} or the set of rational numbers. Instead of continuing such an enumeration of nonhomeomorphic sets, we content ourselves with mentioning that they are more abundant than their homeomorphic counterparts.

A function $f:X \to Y$ is <u>open</u> iff $f(U)$ is open in Y for each open set $U \subseteq X$. It is easy to see that a function h is a homeomorphism if and only if h is 1-1, continuous and open.

For future use we mention here that a function $f:X \to Y$ which has the property that $f(X) = Y$ is said to be <u>onto</u>, or synonymously, <u>surjective</u>.

Section 5. Product spaces

It is convenient to consider product spaces in two separate categories: (1) those with a finite number of member spaces; and (2) those with an infinite number of member spaces. Actually, product spaces of type (2) will not be of use in the following chapters on semigroups; nevertheless, because of their importance for theorems that we will use, such as Tychonoff's product theorem for compact spaces and a corresponding result for connected spaces, infinite product spaces will be briefly introduced here.

A collection $\mathcal{B}$ of subsets of a space $(X, \mathcal{T})$ is called a <u>base</u> and said to "generate" $\mathcal{T}$ iff every member of $\mathcal{T}$ can be written as a union of members of $\mathcal{B}$. In this regard a topology is characterized by a base, but it is noted that a base need not be unique.

You are no doubt used to considering the plane in cartesian form as $R \times R$; that is, each point is an ordered pair (a,b) and similarly points in 3-dimensional space are usually designated by ordered triples with respect to three mutually perpendicular axes, each axis being a real line. These are examples of finite product spaces in which every member of the product spaces is the same, namely, R. One might be tempted to extend this notion so that a product space with n member spaces may be thought of as a collection of n-tuples, $x = (x_1, x_2, \ldots, x_n)$. Yet it is more precise, and often more useful, to consider these cartesian products as collections of functions as follows:

Let k be a fixed integer and let N_k denote the set of natural numbers, 1, 2, ..., k. The set of all functions x from N_k into $\cup\{X_i;\ i = 1,\ \ldots,\ k\}$ such that $x(i) \in X_i$ is called the <u>cartesian product</u> of the spaces X_i and is denoted by $X = X_1 \times X_2 \times \ldots \times X_k = \times\{X_i;\ i \in N_k\}$. The image points $x(i)$ of the function x are usually written x_i. Notice that the plane, according to this definition, is the collection of functions from the set $\{1,2\}$ into $R \cup R = R$.

The collection $\mathcal{B} = \{U_1 \times \ldots \times U_k;\ U_i \in \mathcal{T}_i$ for each $i \in N_k\}$ is a base for a topology for the finite product space $X_1 \times \ldots \times X_k$. This is called the <u>product topology</u>. For each $i \in N_k$ we may define a <u>projection</u> function $\pi_i : X \to X_i$ by $\pi_i(x) = x_i$. It is clear that a projection is surjective. The significance of the product topology is that it is a

14

topology for X which makes all the projection functions continuous. This is easily seen by noting, for $U_i \in \mathscr{T}_i$, that $\pi_i^{-1}(U_i) = X_1 \times X_2 \times \ldots \times U_i \times \ldots \times X_k$ which itself is an open set in $\mathscr{B}$. In addition, it may easily be shown that the product topology is the smallest topology for X which ensures that the projection functions are continuous [6].

Before turning to infinite product spaces, we pause for a moment to establish a simple result we will need in the succeeding chapters. In the proof of this result we will need the fact that a point p is in A* if $A \cap U \neq \phi$ for each open set U which contains p. (Since this fact is an alternative definition for the closure of a set, i.e., $A* = \{p;\ A \cap U \neq \phi$ for each open set U containing p$\}$, we will omit its proof.)

1.3. <u>Proposition</u>. If A and B are subsets of spaces $(X, \mathscr{U})$ and $(Y, \mathscr{V})$, respectively, then $(A \times B)* = A* \times B*$.

Proof. Suppose $(a,b) \in (A \times B)*$ and let $U \in \mathscr{U}$ be any open set containing a. $U \times Y$ is open in $X \times Y$ and, by assumption, $(U \times Y) \cap (A \times B) \neq \phi$. It follows, easily, that $U \cap A \neq \phi$ and so $a \in A*$. Similarly, $b \in B*$ and, consequently, $(A \times B)* \subset A* \times B*$.

If $(a,b) \in A* \times B*$, then $a \in A*$ and $b \in B*$. Any open set W containing (a,b) has the form $\bigcup_\alpha \{U_\alpha \times V_\alpha\}$ where $a \in U_k$ and $b \in V_k$ for some fixed k. $U_k \cap A \neq \phi$ and $V_k \cap B \neq \phi$ so that $(U_k \times V_k) \cap (A \times B) \neq \phi$ Therefore, $W \cap (A \times B) \neq \phi$

and $A^* \times B^* \subset (A \times B)^*$. $\square$

We now turn our attention to infinite product spaces. The Cartesian product of an infinite number of topological spaces may be defined as a collection of functions just as in the finite case with the modification that the index set may be any set, and not N_k for some fixed k. The index set may be the set of natural numbers N, if denumerable, or it may be some uncountable set.

In looking for an appropriate topology, we look for the smallest topology $\mathscr{S}$ for X which ensures that the projection functions are continuous, as in the finite case. It turns out that a base member for $\mathscr{S}$ will be of the form $\times\{U_\alpha;$ $\alpha \in A$, an index set, and U_α in $\mathscr{T}_\alpha\}$ where $U_\alpha = X_\alpha$ for all but a finite number of α's in A. This, of course, is not the immediate generalization, of the finite situation, that one might expect because of the restriction on the components U_α. However, it is an extension of the product topology for the finite case in that if A is a finite index set the two topologies coincide. For this reason $\mathscr{S}$ also is called the product topology.

We have mentioned that projection functions are both onto and continuous functions. Additionally they are open maps:

1.4. <u>Proposition</u>. Each projection map of a product space is an open function.

Proof. Let $V = \times\{U_a;\ a \in A\}$ be a member of the base for the topology $\mathcal{S}$. Recall that $U_a = X_a$ for all but a finite number of a's. Thus, for any $a \in A$, $\pi_a(V) = U_a$, which may be X_a, and so is open in X_a. If W is any open set in the product space X, then W is the union of a family $\mathcal{F}$ of base members of the form V. Hence, $\pi_a(W) = \pi_a(\cup \mathcal{F}) = \cup\{\pi_a(F)$; $F \in \mathcal{F}\}$ and we have shown that each member of the last union is open. Therefore, π_a is open. $\square$

It is natural to ask whether projections are also closed; i.e., $\pi(C)$ is <u>closed</u> for each closed set C contained in the product space X. This is not the case as shown, easily, by counterexample [6].

We will end this section with two results concerning product spaces when the component spaces are Hausdorff:

1.5. <u>Theorem</u>. The product of Hausdorff spaces is Hausdorff.

Proof. Consider two distinct points x and x' in the product space $X = \times\{X_\lambda$; $\lambda \in A$, an index set$\}$. Since $x \neq x'$, there is at least one $\alpha \in A$ such that $x_\alpha \neq x'_\alpha$. Then there exist disjoint open sets U and V in X_α containing x_α and x'_α, respectively, because X_α is Hausdorff. $\pi_\alpha^{-1}(U)$ and $\pi_\alpha^{-1}(V)$ are open, disjoint sets containing x and x', respectively. They are open because π_α is continuous and they are disjoint because if b is in their intersection, then $\pi_\alpha(b) \in U \cap V$, a contradiction. $\square$

For the next result we introduce one new concept: Let Y be any topological space. Then $\{(y, y)$; $y \in Y\}$ is called the <u>diagonal</u> of Y × Y and is often denoted by $\Delta(Y)$, or simply Δ if Y is clearly understood to be the appropriate space.

    1.6. <u>Proposition</u>. If Y is a Hausdorff space, then Δ is closed.

    Proof. We will show that no element in $(Y \times Y) \setminus \Delta$ is in Δ^*. If y and y' are distinct elements of Y, then there are disjoint open sets U and V containing y and y', respectively. $(y,y') \; \varepsilon \; U \times V$ which is an open set in the product topology. If, for any $z \; \varepsilon \; Y$, $(z, z) \; \varepsilon \; U \times V$ we would have $z \; \varepsilon \; U \cap V$, a contradiction. Hence $\Delta \cap (U \times V) = \phi$ and so $(y,y') \notin \Delta^*$. $\square$

    The converse of (1.6), namely, "if Δ of X × X is closed, then X is Hausdorff" is also true. However, since we will always be dealing with Hausdorff spaces in the chapters that follow, the converse will be of no use to us and so we omit its proof.

Section 6. Compact spaces

    The notion of a compact space will play an important role in our study of topological semigroups. Moreover, it occupies an eminent position in the development of topology itself.

    Let $(X, \mathcal{T})$ be a topological space. A collection $\mathcal{U}$ of members of $\mathcal{T}$ is an open <u>cover</u> for a subset A of X iff $A \subseteq \cup \{U; \; U \; \varepsilon \; \mathcal{U}\}$. $\mathcal{U}$ is a <u>finite cover</u> iff $\mathcal{U}$ has a finite number of members. $\mathcal{V}$ is a <u>subcover</u> of $\mathcal{U}$ iff $V \; \varepsilon \; \mathcal{U}$ for every member V in $\mathcal{V}$ and $\mathcal{V}$ covers the space. A space X

18

is <u>compact</u> iff every open cover of X has a finite subcover. A subset A of X is <u>compact</u> iff every open cover of X has a finite subcover. A subset A of X is compact iff A is a compact space with the relative topology. It follows easily that a subset A, of a space $(X, \mathcal{T})$, is compact iff every open cover of A consisting of members of $\mathcal{T}$ contains a finite subcover of A. This last formulation of compactness of a subset is the one we will use.

The idea of a compact set is an abstraction of a property possessed by certain subsets of the real numbers. Worthy of mention is the Heine-Borel Theorem for the set of real numbers, R, which characterizes the compact subsets of R as being precisely those subsets A which are closed and bounded, i.e., bounded in the sense that there exist positive real numbers r and s such that $-r < a < s$ for every element $a \in A$). Thus all closed intervals, including single points, are compact. Since R itself is not compact, a concept called local compactness is used to describe the properties of R in this area. Local compactness, though very important, will not be treated here.

We proceed with a result relating closed sets to compactness:

1.7 <u>Proposition</u>. A closed sub set of a compact space is compact.

Proof. Let A be a closed subset of a compact space X and $\mathcal{U}$ be an open cover of A. Consider the open cover $\mathcal{V}$ of X consisting of X\A and all U in $\mathcal{U}$. $\mathcal{V}$ has at least one finite subcover for X, one being $\bigcup_{i=1}^{n} U_i \cup X \backslash A$ for some positive integer n. Since $A \cap X\backslash A = \phi$, $A \subset \bigcup_{i=1}^{n} U_i$ and so $\mathcal{U}$ has a finite subcover of A.$\square$

The converse of this result, namely, "A compact subset of a compact space is closed" is not true. This may easily be seen by considering any set with more than one element and selecting the indiscrete topology. Fortunately, such a space is not Hausdorff and Hausdorff spaces are our concern in this book.

1.8. <u>Proposition</u>. A compact subset of a Hausdorff space is closed.

Proof. Let C be a compact set in a space X and select any point p in X\C. Since X is Hausdorff, to each point $c \in C$ there correspond two disjoint sets U_c and V_c containing c and p, respectively. The collection $\mathcal{U} = \{U_c;\ c \in C\}$ is an open cover of C and so there exists a finite subcover, say, $C \subset \bigcup_{i=1}^{n} \{U_{ci};\ U_{ci} \in \mathcal{U}\}$ for some positive integer n. Consider the set $\bigcap_{i=1}^{n} V_{ci}$ which is open, contains p and does not intersect C. Therefore, $p \notin C^*$ and so $C = C^*$.$\square$

One of the most significant theorems in topology is the Tychonoff Product Theorem, so named after the Russian mathematician-geophysicist Andrey Nikolaevich Tikhonov, which deals with the compactness of a product space. This result is an important consequence of the manner in which we defined the product topology for an infinite number of spaces and, for this reason, the product topology is often referred to as the Tychonoff topology. At this point the reader may choose to read only the statement of (1.10) and to omit the proof which is somewhat involved and is not necessary for an understanding of topological semigroups. In fact, only a sketch of the proof is presented here; further discussion is found in [2] and alternative proofs are given in [3] and [6]. It is noted that the axiom of choice, in one of its many equivalent forms, is necessary to establish the Tychonoff Product Theorem; in fact, Kelley [4] has shown that this theorem itself is equivalent to the axiom of choice! The method of proof we will sketch rests upon the Hausdorff Maximal Principle and an equivalent formulation of compactness:

A family $\mathscr{F}$ of sets, in any space, is said to have the _finite intersection property_, abbreviated FIP, iff each finite subcollection of $\mathscr{F}$ has a nonempty intersection.

1.9. <u>Proposition</u>. A space is compact if and only if each family of closed sets which has the FIP has itself a nonempty intersection.

Proof. Both parts are proved by contradiction, making use of De Morgan's Laws. Let X be a compact space and $\mathscr{F} = \{C_\alpha ; \alpha \in A$, an index set$\}$ be a collection of closed sets with FIP such that $\cap_\alpha C_\alpha = \phi$. This implies that each point $x \in X$ is in $X \backslash C_\alpha$ for some α. Thus $\{X \backslash C_\alpha ; \alpha \in A\}$ is an open cover of X and so $\bigcup_{i=1}^{n} \{X \backslash C_i\} = X$ for some positive integer n. Therefore, we obtain $X \backslash \bigcap_{i=1}^{n} C_i = X$ which implies that $\bigcap_{i=1}^{n} C_i = \phi$, contradicting the fact that $\mathscr{F}$ has the FIP.

Conversely, suppose each family of closed sets with FIP has a nonempty intersection and yet X is not compact. Let $\mathscr{U} = \{U_\alpha;\ \alpha \in A,$ an index set$\}$ be an open cover of X which has no finite subcover; i.e., for all n, $X \backslash \bigcup_{i=1}^{n} U_i \neq \phi$. By one of De Morgan's Laws, $\bigcap_{i=1}^{n} X \backslash U_i \neq \phi$ for all n. Therefore, by hypothesis, $\{X \backslash U;\ U \in \mathscr{U}\}$ has a nonempty intersection. Consequently, $\mathscr{U}$ is not a cover of X, contrary to assumption. $\square$

1.10. <u>Tychonoff Product Theorem</u>. The cartesian product of a collection of compact topological spaces is compact with respect to the product topology.

Proof. Let $\mathscr{F} = \{C_\gamma;\ \gamma \in F,$ an index set$\}$ be a collection of closed sets with FIP in the product space $X = \times\{X_\alpha;\ \alpha \in A,$ an index set$\}$. By the Hausdorff Maximal Principle we can find a collection $\mathscr{G} = \{D_\beta;\ \beta \in B,$ an index set$\}$ of subsets of X with the properties:

(i) $\mathscr{F}$ is a subcollection of $\mathscr{G}$;

(ii) $\mathscr{G}$ has the FIP:

(iii) $\mathscr{G}$ is maximal with respect to (i) and (ii), i.e., $\mathscr{G}$ is not a proper subcollection of any other collection of sets having properties (i) and (ii);

(iv)   the intersection of any finite subcollection of $\mathscr{G}$ is itself a member of $\mathscr{G}$; and

(v)   any set in X that intersects every member D_β in $\mathscr{G}$ is itself a member of $\mathscr{G}$.

For each $\alpha \in A$, the collection $\{\pi_\alpha(D_\beta); \beta \in B\}$ has the FIP. However, these sets are not necessarily closed. The collection $\{[\pi_\alpha(D_\beta)]^*; \beta \in B\}$ still has the FIP and hence, since X_α is compact, there is a point x_α common to each set $[\pi_\alpha(D_\beta)]^*$. In X let x be the point whose coordinates are the points x_α and let $V = \times\{V_\alpha; \alpha \in A\}$ be a base member containing x. Because of the nature of the product topology and property (iv) of $\mathcal{G}$, V is a member of $\mathcal{G}$. Therefore, since $\mathcal{G}$ has the FIP, $V \cap C_\gamma \neq \phi$ for each $\gamma \in F$. Since V was an arbitrary base member containing x, $x \in C_\gamma^*$ for all γ and, thus, $x \in C_\gamma$ for all γ because $\mathcal{F}$ is a collection of closed sets. Therefore, $\mathcal{F}$ has a nonempty intersection. $\square$

We conclude this section with two theorems relating continuous functions and compact sets:

1.11. **Theorem.** If A is a compact subset of a space X and f is a continuous function defined on X, then f(A) is compact.

Proof. Let $\mathcal{V}$ be any open cover of f(A). $\{f^{-1}(V); V \in \mathcal{V}\}$ is an open cover for A and, since A is compact, there is a natural number n such that $\bigcup_{i=1}^{n}\{f^{-1}(V_i); V_i \in \mathcal{V}, i = 1, 2, \ldots, n\}$ covers A. Since $\bigcup_{i=1}^{n} f^{-1}(V_i) = f^{-1}[\bigcup_{i=1}^{n} V_i]$, $f(A) \subseteq \bigcup_{i=1}^{n} V_i$ and, therefore, f(A) is compact. $\square$

This result takes on special significance when the function is a homeomorphism and the set A is the space X. In this case the theorem tells us that compactness is a topological invariant.

The next theorem, named in honor of the eminent topologist, Alexander Doniphan Wallace, seems simple enough; however, its consequences are numerous and its proof is nontrivial. In fact, the reader is encouraged to omit the proof if, after an honest effort, it becomes conceptually too difficult. For, the techniques involved in the proof will not be of particular interest in topological semigroups.

1.12. <u>Wallace Theorem</u>. If A and B are compact subsets of spaces X and Y, respectively, and if W is an open set containing $A \times B$ in $X \times Y$, then there exist open sets U and V such that $A \subset U$, $B \subset V$ and $U \times V \subset W$.

Proof. Fix an element a in A. For each y in B, there exist open sets J of a and K of y such that $J \times K \subset W$. $\{K_y; y \in B\}$ cover B, which is compact, so that a finite number do, say, K_{y1}, K_{y2}, ..., K_{yn} for some natural number n. To each K_{yi} there is a corresponding J_i and $a \in P_a = \bigcap_{i=1}^{n} \{J_i; i = 1, 2, ..., n\}$. Then, letting $Q = \bigcup \{K_{yi}; i = 1, ..., n\}$ we find $B \subset Q$ and $P_a \times Q \subset W$.

The collection $\{P_a; a \in A\}$ is an open cover for A and so a finite subcollection covers A, say, $A \subset U = \bigcup_{i=1}^{m} P_{ai}$ for some natural number m. To each P_{ai} there corresponds a Q_i and, clearly, $B \subset V = \bigcap_{i=1}^{m} Q_i$. Since $V \subset Q_i$ for all $i = 1, 2, ..., m$, it is immediate that $P_{ai} \times V \subset W$, for the same values of i, and so $\bigcup_{i=1}^{m} (P_{ai} \times V) \subset W$. Therefore,

U × V ⊆ W (by viewing the 1.22 Exercise in this chapter). $\square$

It is easy to relate continuity to the Wallace theorem:

1.13. <u>Corollary</u>. Let A and B be compact subsets of spaces X and Y, respectively, and let f:X × Y → Z, another space, be a continuous function. If W is an open subset containing f(A × B), then there exist open sets U and V such that A ⊆ U, B ⊆ V and f(U × V) ⊆ W.

Proof. Since f is continuous, f^{-1}(W) is open in X × Y. Letting J = f^{-1}(W), it is clear that A × B ⊆ J. By the Wallace Theorem there exist open sets U and V such that A ⊆ U, B ⊆ V and U × V ⊆ J. Consequently, f(A × B) ⊆ f(U × V) ⊆ f(J) ⊆ W. $\square$

This section ends with a result which is closely related to the Wallace Theorem. Indeed, although it may be established independently, the proof we have chosen to present here is an elegant and simple consequence of the Wallace Theorem. This is appropriate because in the semigroup chapters both corollaries will be utilized together. We first introduce the notion of a normal space:

A space (X, $\mathcal{T}$) is <u>normal</u> iff for each pair of disjoint closed sets A and B in X there exist disjoint open sets U and V in $\mathcal{T}$ such that A ⊆ U and B ⊆ V.

1.14. <u>Corollary</u>. A compact Hausdorff space is normal.

Proof. Let X be a compact Hausdorff space and A, B be disjoint closed sets of X. X is Hausdorff implies (X × X)\Δ is open and this set contains A × B because of their disjoint-

ness. Also, A and B are compact because X is compact.
Thus, by the Wallace Theorem, there are open sets U, V
such that $A \subset U$, $B \subset V$ and $U \times V \subset (X \times X)\backslash\Delta$. In this
instance $U \cap V = \phi$, for otherwise, if z is in $U \cap V$,
then $(z,z) \in U \times V \subset (X \times X)\backslash\Delta$ which is clearly, a contradiction.□

Section 7. Connected spaces

The intuitive notion of connectivity may be expressed
in an abstract topological manner. There are various types
of connectedness. For example, consider a blackboard in
contrast to a doughnut. A connected object with no "holes"
is said to be <u>simply connected</u>. However, it will not be
our task here to study types of connectedness [1]; and the
results we obtain will apply to all types. We proceed in
a formal fashion:

Two subsets A and B of a space are <u>separated</u> iff
$A \cap B^* = \phi$ and $A^* \cap B = \phi$. A topological space X is <u>connected</u>
iff X is not the union of two nonempty separated sets. A
subset Y of X is connected iff the space Y with the relative
topology is connected.

There are three results on connectedness which will
be of particular use in semigroups. For that reason some
preliminary ideas will be termed lemmas, although they are
of interest themselves.

Because the definition of connectivity is of a

negative nature, it is not surprising that a number of the proofs in this section involve the logically equivalent contrapositive or the indirect method of contradiction.

1.15. <u>Lemma</u>. Let A, B and C be nonempty subsets of a space X. If A and B are separated, if C is connected, and if $C \subset A \cup B$, then $C \subset A$ or $C \subset B$.

Proof. Assume the conclusion is false; namely, both the sets $C \cap A$ and $C \cap B$ are nonempty. For notation let $D = C \cap A$ and $F = C \cap B$; note that $D \cup F = C$. Clearly, $D^* \subset A^*$ and so $D^* \cap F \subset A^* \cap B$ which is empty since A and B are separated. In the same manner $F^* \cap D = \phi$ and, therefore, D and F are separated sets. But this contradicts the fact that C is connected. Consequently, our assumption was incorrect so that either $C \subset A$ or $C \subset B$. $\square$

1.16. <u>Proposition</u>. Let A be a connected subset of a space X and $\mathscr{C}$ be a family of connected subsets of X. If each member of $\mathscr{C}$ is not separated from A, then $A \cup (\cup \mathscr{C})$ is connected.

For reasons of intuition, it may be beneficial to consider a possible pictorial view of this proposition:

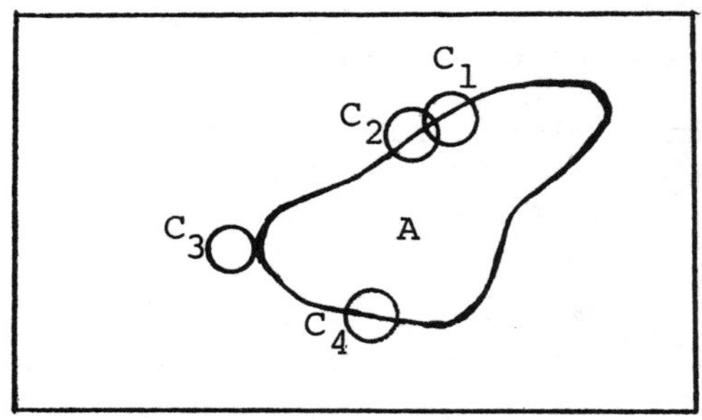

It is noted that there is no reason for the members of $\mathcal{C}$ to be connected to each other. Now let us proceed with the proof:

Proof. We will establish the contrapositive; namely, if A ∪ (∪ $\mathcal{C}$) is not connected, then A and at least one C ε $\mathcal{C}$ are separated. Since A ∪ (∪ $\mathcal{C}$) is not connected, it is the union of two nonempty separated sets D and F. By (1.15) A is in D or F; the same is true for each C in ∪ $\mathcal{C}$. Suppose A is in D. Therefore, since F ≠ φ, there is a C' ε $\mathcal{C}$ which is in F. It follows, easily, that A and C' are separated, proving the contrapositive. □

We turn our attention now to a theorem implying that connectedness is a homeomorphic property. As preparatory material we give equivalent formulations of separated sets and connectedness:

1.17. <u>Lemma</u>. Let A and B be disjoint subsets of a space X. A and B are separated if and only if A and B are both open and closed in A ∪ B.

Proof. It is evident that A is open in A ∪ B if and only if B is closed in A ∪ B. Thus it suffices to show that A and B are separated if and only if A and B are both open in A ∪ B. First suppose that A and B are separated. In the relative topology the set (X\A*) ∩ (A ∪ B) = B\A* is open in A ∪ B. Since A* ∩ B is empty, B\A* = B. In a similar fashion A ∩ B* = φ implies that A is open in A ∪ B.

Conversely, if A is open in A ∪ B, then B = (A ∪ B)\A

28

is closed in $A \cup B$; that is, $A \cap B^* = \phi$. Similarly, B
being open in $A \cup B$ implies $A^* \cap B = \phi$ and so A and B
are separated.$\square$

1.18. <u>Lemma</u>. A space X is connected if and only
if it has no proper subset which is both open and closed.

Proof. Suppose there is no proper subset which is both
open and closed. Then X cannot be the union of two separated
sets by (1.17).

To prove the converse we consider its contrapositive.
Suppose A is a proper subset which is open and closed in X.
Then X\A is open and closed in X also. By (1.17) A and X\A
are separated and so X is not connected.$\square$

1.19. <u>Theorem</u>. If A is a connected subset of a space
X and if $f: X \to Y$ is a continuous function, then $f(A)$ is
connected.

Proof. Consider A and $f(A)$ to be spaces with the
relative topologies of X and Y, respectively, and let g
be a function from A to Y defined by $g(x) = f(x)$ for all
x in A. (g is called a <u>restriction</u> of f.) It is easy to
verify that g is continuous if f is continuous and, clearly,
$f(A) = g(A)$. We will establish the contrapositive. Suppose
that f is continuous and $g(A)$ is not connected. By (1.18)
there is a proper subset B of $g(A)$ which is both open and
closed in the space $g(A)$. Since g is continuous, $g^{-1}(B)$
is both open and closed. $g^{-1}(B)$ is a proper subset of A
and so, again using (1.18), A is not connected.$\square$

The "size" of a connected set is our next topic. A component is a maximal connected subset of a space, that is, a connected subset which is not properly contained in any other connected subset. If a space has the discrete topology, then each point constitutes a component; in contrast, any connected space, and in particular the space of real numbers, has no proper subset as a component. As we will now see, components have the important property that all of them are closed sets, and, of equal significance, they are disjoint and generate a "connectedness" equivalence relation, partitioning any space.

1.20. Lemma. If A is a connected subset of a space X and if $A \subseteq B \subseteq A^*$, then B is connected.

Proof. Assume that B is not connected; i.e., $B = D \cup F$ where D and F are nonempty separated sets. By (1.15) A is a subset of D or F, say $A \subseteq D$. Since $F \subseteq A^*$, F is also contained in D^* contradicting the fact that D and F are separated. Hence the assumption that B is not connected is incorrect. $\square$

The case when $B = A^*$ in (1.20) is worthy of comment. For, if a connected set is not closed, this says that it can be "enlarged" to a closed set and yet retain its connectivity.

1.21. Proposition. Each component of a space is closed and no two components intersect.

Proof. If C is a component which is not closed, then

C* is connected, by the remark preceding this proposition, and it contains C, contradicting the maximality of C. Therefore, C is closed.

Suppose C and C' are two distinct intersecting components. Letting A be any singleton subset of $C \cap C'$ and $\mathscr{C}$ = {C,C'}, (1.16) states that $C \cup C'$ is connected, contradicting the maximality of both C and C'. Thus no two components inter-sect. $\square$

We mention without proof that connectedness has a counterpart to the Tychonoff Product Theorem, namely, the product of any number of connected spaces is a connected (product) space. The product space composed of two connected spaces, which is of special interest for semigroups, is connected by a particularly simple proof which is based on (1.16) and may be found in [6].

EXERCISES

1.22. (a) Use an "element argument" to prove the set-theoretic result, $(\bigcup_{\alpha} B_{\alpha}) \times D = \bigcup_{\alpha} (B_{\alpha} \times D)$ where $\alpha \in A$, an index set.

(b) Use the plane, i.e., $R \times R$, to illustrate that $(B_1 \cup B_2) \times (D_1 \cup D_2)$ need not equal $(B_1 \times D_1) \cup (B_2 \times D_2)$.

1.23. Prove that a subset of a Hausdorff space is Hausdorff.

1.24. Let $f: X \to Y$ be an onto function; i.e., $f(X) = Y$. What is the smallest topology for Y which will ensure that f is continuous?

1.25. Prove that the restriction of a continuous function is continuous.

1.26. Let X be a compact space and Y be Hausdorff. Prove that if A is a subset of X and $f: X \to Y$ is continuous, then $f(A^*) = f(A)^*$.

1.27. Prove that if $f: (X, \mathcal{T}) \to (Y, \mathcal{U})$ is a function which is onto, continuous and either open or closed, then $\mathcal{U}$ is the quotient topology.

1.28. Let X be a compact space and Y be Hausdorff. Prove that if $f: X \to Y$ is onto, 1-1 and continuous, then X and Y are homeomorphic. (The "reverse" situation is not true, that is, a 1-1 continuous function of a Hausdorff space onto a compact space need not be a homeomorphism. For example, let $X = Y =$ the unit interval $[0,1]$, $\mathcal{D}$ be the

32

discrete topology, $\mathcal{T}$ be the relative topology with respect to the real numbers and $f:(X,\mathcal{D}) \to (Y,\mathcal{T})$ be the identity map.)

1.29. (a) Prove or disprove: The intersection of two connected sets is connected.

(b) Relate your result in (a) to the proof of (1.21).

1.30. Prove that a set U is open if and only if for each point x in U there exists an open set V_x containing x with the property that $V_x \subset U$. (This result is a useful criteria sometimes for proving that a set is open.)

1.31. Let f be a function with domain X. Prove that f is continuous if and only if for each x in X and each open set V containing f(x) there exists an open set U containing x with the property $f(U) \subset V$.

1.32. Let R be the set of all real numbers and let k be any fixed real number. Prove that the following two functions are continuous:

(a) the constant function $c:R \to R$ defined by c(x) = k;

(b) the identity function $e:R \to R$ defined by e(x) = x.

1.33. Find a counterexample to disprove the following statement: The continuous image of a disconnected set is disconnected.

REFERENCES FOR CHAPTER I

1. Arnold, B.H., *Intuitive Concepts in Elementary Topology*. Englewood Cliffs, N. J.: Prentice-Hall, 1962.

2. Hocking, J.G., and Young, G.S., *Topology*. Reading, Mass.: Addison-Wesley, 1961.

3. Kelley, J.L., *General Topology*. Princeton, N. J.: Van Nostrand, 1955.

4. ______________, "The Tychonoff product theorem implies the axiom of choice," Fund. Math. 37, 75-76 (1950).

5. Lietzmann, W., *Visual Topology*. New York: American Elsevier Publishing Co., 1965.

6. Moore, T.O., *Elementary General Topology*. Englewood Cliffs, N. J.: Prentice-Hall, 1964.

7. Steenrod, N.E., and Chinn, W.G., *First Concepts of Topology*. New York City: Random House and L. W. Singer Co., 1966.

CHAPTER II.   TOPOLOGICAL SEMIGROUPS

Section 1.   Introductory Concepts

A (topological) <u>semigroup</u> is a nonnull Hausdorff space
S together with a function $m: S \times S \to S$ such that m is continuous and associative. Any algebraic semigroup can become a
(topological) semigroup by adding a discrete topology to
its structure. For any a, b ε S, let m(a,b) = ab. That
is, the function m --called multiplication-- is denoted by
juxtaposition. Also, for subsets, A, B of S, define AB =
$\cup$ {ab; a ε A, b ε B}. Note that AB = m(A $\times$ B).

We pause for a moment to enumerate a number of algebraic
and topological entities which motivate a study of semigroups.
For notation, letting a be any real number and B be a subset of real numbers define B(a) to be the elements in set
B which are $\geq$ a. In addition, let R be the set of real
numbers, R* be the rational numbers, I be the closed interval
[0,1] of real numbers, Z be the integers, E be the even integers and D be the odd integers; note that all these sets are
infinite and all, except for I, contain negative numbers.
We are now ready to determine a number of subsets of the real
numbers which form semigroups. Under addition we have the
following: (1) R, R*, Z, E;   (2) R(a), R*(a), Z(a) and
E(a) for any a $\geq$ 0 --one of these worthy of special mention
is Z(1) which is sometimes called the set of natural numbers;
(3) the singleton subset {0}. Under multiplication we also

have the following semigroups: (a) R, R*, Z, E, D, I; (b) R(a) and R*(a) for a = 0 and any a $\geq$ 1; (c) Z(a), E(a) and D(a) for any a $\geq$ 0; (d) the finite subsets {0}, {1}, {0,1}, {-1,1}, {-1,0,1}; (e) the interval [-1,1]. It is remarked that these do not exhaust the subsets of real numbers which are semigroups under usual multiplication and addition. For example, fix any real number r and consider all its integer multiples, both negative and nonnegative, under either operation; of course, Z and E are special types of this kind of semigroup with r = 1 and 2, respectively. Also, these semigroups contain other semigroups within them; for instance, consider the positive or nonnegative multiples of r $\geq$ 0. All the above examples become topological semigroups if we impose the discrete topology. However, using the topology called the "usual" topology for real numbers, which we discussed in Chapter 1, the sets R, R(a), I and the interval [-1,1] take on more significance topologically.

An element e_L in a semigroup S is a <u>left identity</u> of S iff $e_L s = s$ for each s ε S. An element o_L in S is a <u>left zero</u> of S iff $o_L s = o_L$ for each s ε S. An element e which is both a left and right identity is called an <u>identity</u> (or unit); a <u>zero</u> is similarly defined.

2.1. (algebraic) <u>Proposition</u>. If a semigroup has both a left and a right identity (zero), then they are the same element and this identity (zero) is unique.

In other words, (2.1) says that a semigroup possesses at most one identity and at most one zero. The proof is left for the reader to do as an exercise.

Section 2. Subsemigroups and Idempotents

If A is a subset of a semigroup S, then A is a <u>subsemigroup</u> of S iff $A^2 \subset A$. In one of the proofs we will use the concept of a group: A semigroup S is an <u>abstract group</u>, or simply <u>group</u>, iff aS = Sa = S for all a ε S.

2.2. (algebraic) <u>Proposition</u>. If a semigroup S contains more than one element, then S contains a subsemigroup S' such that S' $\neq$ S.

Proof. Suppose that each such S' = S. Note that, for any a ε S, $(aS)^2$ = (aS)(aS) = a(SaS) $\subset$ aS. Thus, by assumption, Sa = S = aS and so S is a group. e, the identity of S, constitutes a semigroup and {e} $\neq$ S, a contradiction. $\square$

2.3. <u>Lemma</u>. If A, B $\subset$ S, then A*B* $\subset$ (AB)*; if, in addition, A* and B* are compact, then A*B* = (AB)*.

Proof. A*B* $\equiv$ m(A* $\times$ B*) = m((A $\times$ B)*)$\subset$[m(A $\times$ B)]* = (AB)*.

If, in addition, A* and B* are compact, then m(A* $\times$ B*) = A*B* is closed (because, among other things, a compact set in a Hausdorff space is closed as shown in (1.8)). Therefore, (AB)* $\subset$ (A*B*)* = A*B*. $\square$

2.4. <u>Proposition</u>. If A is a subsemigroup of S, then A* is also a subsemigroup.

Proof. Using (2.3) we have $(A*)(A*) \subset (A^2)* \subset A*$. $\square$

The next result is purely topological, but it is an important tool in establishing the necessary continuity for many semigroup multiplications. We will encounter its usage a number of times in this book.

2.5. <u>Lemma</u>. If for each $\lambda \in \Lambda$, an indexing set, we are given a continuous function $f_\lambda : X \to Y_\lambda$, then we may define a continuous function $f : X \to \times \{Y_\lambda ; \lambda \in \Lambda\}$ such that $f_\lambda = \pi_\lambda f$ for all $\lambda \in \Lambda$, that is, $f(x) = (f_1(x), f_2(x), \ldots)$ if the indexing set is taken to be the natural numbers.

Proof. Define $f(x) = y$ iff $\pi_\lambda(y) = f_\lambda(x)$ for all λ. (It is easy to show that $\bigcap_\lambda \pi_\lambda^{-1} f_\lambda(x) = f(x)$. To show the continuity of f, let $\mathcal{U}$ be a base for $\times_\lambda Y_\lambda$. Each open set U in $\mathcal{U}$ has the form $U = \bigcap_{i=1}^n \pi_{\lambda_i}^{-1}(V_i)$ where V_i is an open set in Y_λ. Since $f^{-1}(U) = \bigcap_{i=1}^n f_{\lambda_i}^{-1}(U_i)$, it follows that $f^{-1}(U)$ is open. $\square$

An <u>idempotent</u> is an element $e \in S$ with the property $e^2 = e$. An identity and a zero are examples of idempotents. Result (2.6) states that, collectively, idempotents possess an important topological property. Two proofs are presented because of their contrasting styles. The first method is indirect and, in my opinion, uninspiring; on the other hand, the second approach is direct and by its elegance inspires creativity.

2.6. <u>Proposition</u>. The set of idempotents E of a semi-

38

group is closed.

1st Proof. Suppose $x \in E^*$ and $x^2 \neq x$. Since S is Hausdorff there exist open sets U and W such that $x \in U$, $x^2 \in W$ and $U \cap W = \phi$. $m^{-1}(W)$ is open and contains (x,x). In $m^{-1}(W)$ pick any (open) neighborhood $A \times B$ containing (x,x). Let $V = A \cap B \cap U$ and note that both $x \in V$ and $m(V \times V) \subseteq W$, that is, $V^2 \subseteq W$. Consequently, $V^2 \cap V = \phi$. Now, since $x \in E^*$ there is an $e \in E \cap V$ and so $e = e^2 \in V^2 \cap V$, a contradiction. Therefore, $E = E^*$. $\square$

2nd Proof. Let Y be a Hausdorff space. If $f: X \to Y$ and $g: X \to Y$ are both continuous functions, we may define a function $f \otimes g: X \to Y \times Y$ by $(f \otimes g)(x) = (f(x), g(x))$ and such a function is continuous by (2.5). Since Y is Hausdorff, $\Delta(Y \times Y)$ is closed and so $(f \otimes g)^{-1}(\Delta(Y \times Y)) = \{x;\ f(x) = g(x)\}$ is closed. Letting $f(x) = x^2$ and $g(x) = x$, our result follows. $\square$

Section 3. Groups

In Section 2 we introduced the notion of an algebraic group because it was needed in the proof of (2.2). In this section we single out the group concept itself for consideration. First, several simple examples are given to indicate that groups and semigroups are very distinct algebraic entities. Then, this section begins in earnest with an investigation of topological aspects of groups. Indeed, such an investigation is appropriate at this point because the study of topological groups preceded that of semigroups in the development of topological algebra. However, it is noted that the techniques used to study the topological nature of groups and semigroups differ greatly. This difference in methodology may in a large part be due to the unique topological properties possessed by the identity element of a group. After completing this book for background, the reader who wishes further information on topological groups should consult Lev S. Pontryagin's classic book on the subject [15].

Semigroups need not be groups and, in fact, may have no subgroups. For example, the open interval $I = (0,1)$ with the relative topology and usual multiplication and the set N of all positive integers with the discrete topology under addition are both semigroups having no subgroups. On the other hand, it is also easy to find semigroups which are not groups and yet have subgroups: Consider the set S of nonnegative real numbers under multiplication. The positive reals, R^+, is a proper subgroup of S. For another example let the Boolean semi-group $B = \{0,1\}$ have a multiplication defined by the table

x	0	1
0	0	0
1	0	1

Clearly, B is not a group and both $\{0\}$ and $\{1\}$ are subgroups. B is an instance of the fact that each compact semigroup,

which is not a group, will have the property that each of
its idempotents will be a distinct subgroup.

We now turn our attention to topological aspects of
groups and begin with a purely topological lemma. The
importance of (2.7) lies primarily in part (b) which states
a sufficient condition yielding continuity of a function
having a compact range.

2.7. _Lemma_. (a) If X is any space and Y is a compact
space, then the projection $\pi_1:X \times Y \to X$ is a closed function.

(b) Let $f:X \to Y$ and let $G(f) = \{(x,f(x)); x \in X\}$. If
Y is compact and $G(f)$ is closed, then f is continuous.

Proof. (a) Let C be closed in $(X \times Y)$. We shall show,
using (1.30), that $X\backslash\pi_1(C)$ is open: Let $z \in X\backslash\pi_1(C)$.
$\{z\} \times Y \subseteq (X \times Y)\backslash C$. By Wallace's Theorem there is an open
set $U \subseteq X$ such that $z \in U$ and $U \times Y \subseteq (X \times Y)\backslash C$. $U \cap \pi_1(C) = \phi$
for, if not, let $w \in U \cap \pi_1(C)$ and then $\pi_1^{-1}(w) \cap C \neq \phi$, a
contradiction because $\pi^{-1}(w) \subseteq U \times Y$. Therefore $X\backslash\pi_1(C)$ is
open. $\square$

(b) Let B be closed in Y.
Since $f^{-1}(B) = \pi_1((X \times B) \cap G(f))$, the result easily follows
in view of part (a). (Recall that $(X \times B)^* = X^* \times B^* =
X \times B$.) $\square$

A _topological group_ is a topological semigroup which
is algebraically a group with the additional property that
the inversion function, that is, $x \to x^{-1}$, is continuous.
When referring to a system which is a group in the algebraic

sense, we will follow the common convention of omitting
the term algebraic whenever feasible. However, if such
a system is, in addition, a topological group, we will
avoid ambiguity by specifying the adjective topological.

2.8. <u>Theorem</u>. If a compact semigroup is a group, then
it is a topological group.

Proof. (due to B. J. Pettis). Define $\psi: S \times S \to S \times S$
by $\phi(x,y) = (xy,e)$ where e is the identity. By the (2.5)
Lemma (with $X = S \times S$, $Y_1 = Y_2 = S$) ψ is continuous.
$\Delta(S \times S)$ is closed since S is Hausdorff. $\psi^{-1}(\Delta) =$
$\{(x,y); xy = e\} = \{(x,x^{-1})\}$ is closed and note that it is
the graph of the inversion function (which we will refer
to as i). Since S is compact and G(i) is closed, by the
(b) part of (2.7) we have that i is continuous. $\square$

Examples of a <u>semi-topological</u> group, i.e., a
topological semigroup which is algebraically a group and
not a topological group, will be given in Chapter 7.

2.9. <u>Proposition</u>. If S is compact and G is a group
in S, then G* is a topological group.

Proof. gG = Gg = G for all $g \in G$ so that $G^2 = G$.
G* is compact and thus, by (2.3), $G*G* = (G^2)* = G*$.
Consequently $xG* \subseteq G*$ and $G*x \subseteq G*$ for all $x \in G*$.

Suppose $g \notin xG*$ for some $g \in G$ and some $x \in G*$. S
compact implies $xG*$ is closed and that $S^2 = m(S \times S)$ is
normal (since S^2 is compact and Hausdorff). Consequently,
there exist open sets U and W such that $g \in U$, $xG* \subseteq W$

42

and U ∩ W = φ. By Wallace's Theorem there exist open sets
V and Y such that x ε V, G* ⊂ Y and VY ⊂ W. So, VG* ⊂ W
and g ∉ VG* (for otherwise g ε W, a contradiction). Since
x ε G*, there exists a ε V ∩ G. Finally, note that G = aG
so that g ε aG and then g ε aG ⊂ aG* ⊂ VG*, a contradiction.
Thus G ⊂ xG* for all x ε G* so that G* ⊂ (xG*)* = xG*.

In a similar manner G* ⊂ G*x for all x ε G* and so
G*x = G* = xG* making G* an (algebraic) group. Note that
G* is compact because S is compact. Therefore, by the
(2.8) Theorem, G* is a topological group.□

2.10. (algebraic) <u>Proposition</u>. If e ε E ⊂ S, then
there is a maximal subgroup containing e and distinct
maximal subgroups are disjoint.

Proof. Let G be the set of all g ε S such that ge =
eg = g and such that there exists an element g^{-1} ε S with
the property $gg^{-1} = g^{-1}g = e$ and $g^{-1}e = eg^{-1} = g^{-1}$. Clearly,
G is a maximal subgroup containing e.

Suppose G and G' are maximal subgroups of S and
a ε G ∩ G'. Let e and f be the identities of G and
G' respectively and let ab = e and ac = f. Then e = ab =
(af)b = a(ca)b = (ac)(ab) = fe and f = ca = (ce)a = c(ab)a =
(ca)(ba) = fe so that e = f. Consequently, since G and G'
are maximal, G = G'.□

2.11. <u>Proposition</u>. If S is compact, then each maximal
subgroup of S is closed.

Proof. This result is immediate in view of Propositions
(2.9) and (2.10).□

EXERCISES

2.12. Prove that a set S is a group if and only if for each pair of elements a, b ε S the equations ax = b and ya = b have solutions in S. In your proof, use the definition of a group given in the text. (It is noted that neither this exercise nor the definition of a group given in the text is the "usual" definition of an algebraic group, namely, a semigroup G which possesses an identity e and which has the property that to each element g in G there corresponds an element g' in G such that gg' = g'g = e.)

2.13. Let X be a set and $\mathcal{S}$ the collection of all subsets of X.

(a) Verify that $\mathcal{S}$ is an algebraic semigroup under the operation of set-theoretic union. Is it a group?

(b) Verify that $\mathcal{S}$ is an algebraic semigroup under the operation of set-theoretic intersection. Is it a group?

(c) Discuss the possibilities of making the semigroups in (a) and (b) topological.

2.14. Let X and Y be sets and define a binary operation $\oplus$ as follows: $(x_1, y_1) \oplus (x_2, y_2) = (x_1, y_2)$.

(a) Verify that the system $(X \times Y, \oplus)$ is an algebraic semigroup in which every element is idempotent. (Such a semigroup is called the <u>rectangular band</u> on X × Y. Later, in Chapter 7, we will show that a rectangular band is a topological semigroup.)

44

(b)   A semigroup S is <u>nowhere abelian</u> if ab = ba (a,b ε S) implies a = b. Verify that (X × Y, $\oplus$) is nowhere abelian.

2.15.   Let $\mathcal{N}$ be the collection of all 2 × 2 matrices with positive integer components (i.e., entries). Does $\mathcal{N}$ constitute an algebraic semigroup under ordinary matrix addition? Does it constitute a group?

2.16.   Let ($\mathcal{L}$,×) represent the collection $\mathcal{L}$ of all 2 × 2 matrices with nonzero integer components considered under the operation of ordinary matrix multiplication. Is ($\mathcal{L}$,×) an algebraic semigroup? Is it a group?

2.17.   Let $\mathcal{I}$ be the collection of all 2 × 2 matrices with integer components. Does ($\mathcal{I}$,×) constitute an algebraic semigroup? Does it constitute a group?

2.18.   Let $\mathcal{M}$ be the collection of all 2 × 2 matrices with real numbers as components. Does ($\mathcal{M}$,×) constitute an algebraic semigroup? Is it a group?

2.19.   Let $\mathcal{P}$ be the collection of all 2 × 2 matrices with positive real number components. Is ($\mathcal{P}$,×) an algebraic semigroup? Is it a group?

2.20.   Which of the examples enumerated at the beginning of this chapter are groups?

2.21.   Let B<a> denote the set of elements in B, a subset of R, which are ≤ a, a real number (not necessarily in B). Is R<0> a semigroup under (a) addition? (b) multiplication?

2.22. Let M(r) denote all the multiples of a fixed real number r. What arithmetic axioms and properties are needed to show that M(r) is (a) a semigroup under addition; and (b) a semigroup under multiplication?

2.23. Give an example to show that a maximal subgroup may fail to be closed if the semigroup is not compact. (Hint: Consider R(0).)

a left ideal is closed on left mult. by elem. of S.

CHAPTER III. IDEALS

Section 1. Fundamental Ideas

A nonempty subset L of a semigroup S is a <u>left ideal</u> of S if SL $\subset$ L. A right ideal is defined analogously and, unless otherwise indicated, if a result is proved for left ideals, then a corresponding result exists for right ideals. I is a (two-sided) <u>ideal</u> of S if it is both a right ideal and a left ideal of S.

3.1 <u>Proposition</u>. If L is a left ideal, then L* is a left ideal; the closure of an ideal is itself an ideal.

Proof. If x ϵ S, then xL* = x*L* $\subset$ (xL)* $\subset$ L* by (2.3). That the closure of an ideal is an ideal is immediate since the first part is true also for right ideals.$\square$

If we say L is a left ideal we mean L is a left ideal of S. At times it is of interest to discuss a left ideal L of a subsemigroup T and then we say L is a left ideal of T. Ideals are treated in a similar fashion.

We will continue the convention initiated in the 2.1 Proposition by indicating more than one result by means of parentheses. For example, part (a) of the next result, (3.2), will indicate that the union of any collection of ideals is itself an ideal and that the union of any collection of left ideals is a left ideal.

3.2. (algebraic) <u>Proposition</u>. (a) The union of any collection of ideals (left ideals) is itself an ideal (left ideal).

(b)   If the intersection of a collection of ideals (left ideals) is nonempty, then it is an ideal (left ideal).

(c)   The intersection of two ideals is an ideal.

(d)   If L is a left ideal and R is a right ideal, then $R \cap L$ is nonempty.

Proof.   (a)   Let $\mathcal{U} = \bigcup_{\alpha \varepsilon \mathbf{a}} \{L_\alpha; L_\alpha$ is a left ideal$\}$ where $\mathbf{a}$ is an index set. Then $S\mathcal{U} = S (\bigcup_{\alpha \varepsilon \mathbf{a}} \{L_\alpha\}) = \bigcup_{\alpha \varepsilon \mathbf{a}} SL_\alpha \subset \bigcup_{\alpha \varepsilon \mathbf{a}} L_\alpha = \mathcal{U}$. This is clearly true also for right ideals and thus for ideals.

(b)   Let $\mathcal{N} = \bigcap_{\alpha \varepsilon \mathbf{a}} \{L_\alpha\}$ be such an intersection. $S\mathcal{N} = S (\bigcap_{\alpha \varepsilon \mathbf{a}} L_\alpha) \subset SL_\alpha \subset L_\alpha$ for each $\alpha \varepsilon \mathbf{a}$. Consequently, $S\mathcal{N} \subset \bigcap_{\alpha \varepsilon \mathbf{a}} L_\alpha = \mathcal{N}$. Again this is true for right ideals and thus for ideals.

(c)   Let I and J be any two ideals. $IJ \subset SJ \subset J$ and $IJ \subset IS \subset I$ so that $IJ \subset J \cap I$ and the intersection is nonempty. The result now follows from (b).

(d)   This proof is left as an exercise, namely, (3.17). $\square$

I is a <u>proper</u> ideal of S if I is an ideal and $I \neq S$. In a similar fashion we may define a proper left ideal. We call a semigroup S <u>simple</u> if S does not contain any proper (two-sided) ideals. The next result relates semigroups to groups in terms of types

48   *Def. p. 36*   *semigroup is group iff*
aS = Sa = S for all a ∈ S.

of proper ideals.

3.3. (algebraic) <u>Theorem</u>. A semigroup is a group if
and only if it has no proper left ideals, right ideals and
ideals.

Proof. Suppose S is a group, let I be an ideal in S
and let a ε I. Then $a^{-1}a = e$ ε I and I = S. In a similar
manner, any left or right ideal equals S. (Alternatively,
since S = Sa for any a ε S, we have S = ∪ {Sa; a ε L} =
SL ⊂ L; etc.)

Now suppose S has no proper ideals of any type and
let a ε S. Sa is a left ideal and so S = Sa; likewise
aS is a right ideal and thus S = aS too. □   *since there are*
no proper left ideals

Why?

Section 2. Connectedness and Ideals

One sided ideals may not be connected. The set T =
$\begin{pmatrix} 1 & x \\ 0 & y \end{pmatrix}$; x and y are real numbers satisfying $|x| + |y| \leq 1$
is an algebraic semigroup under ordinary matrix multipli-
cation. If we establish a 1-1 correspondence between T and
a diamond-shaped portion D of the Cartesian plane, namely,
$\begin{pmatrix} 1 & x \\ 0 & y \end{pmatrix}$ ↔ (x,y), then D will be a topological semigroup
in which the "product" of two elements in D is determined
by the corresponding matrix multiplication and the topology
is the relative topology of R × R. The set {(1/2, 0),
(-1/2, 0)} is easily seen to be a left ideal and, clearly,
the two points are disconnected. (See Figure 3.1.)

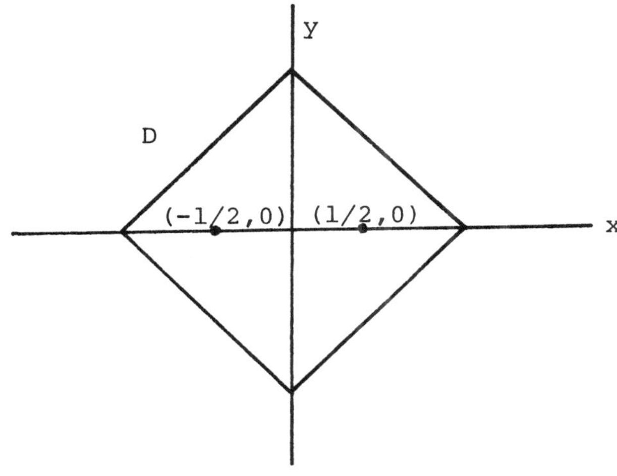

Figure 3.1

Ideals also need not be connected. For example, consider the interval of real numbers S = [0, 1/2], which is a semigroup under multiplication with the usual topology. Let A be any nonempty (not necessarily connected) subset of the open interval (1/4, 1/2). It is easy to see that for any b $\notin$ A, we have Sb $\subseteq$ [0, 1/4] and so S\A becomes an ideal. Clearly, S\A is disconnected because, in par⁻ticular, {1/2} is separated from [0, 1/4] no matter which subset of (1/4, 1/2) is selected as A.

However, as the 3.5 Proposition will show, the presence of an (algebraic) identity ensures topological connectedness of ideals in connected semigroups. First though we state a general result which will be needed in the proof of (3.5):

3.4. <u>Proposition</u>. If A and B are connected subsets of a semigroup, then AB is connected.

The proof of (3.4) is an easy consequence of several topological theorems and the definition of a semigroup. The details are left for the reader as Exercise 3.21.

3.5. <u>Proposition</u>. If a connected semigroup has an identity, then each of its ideals is connected.

Proof. Consider any ideal I of the connected semigroup S. Clearly, $I = \bigcup_{x \varepsilon I} Sx$, one set theoretic inclusion holding since S has an identity and the other being true because I is an ideal. If a is any fixed element of I, then $aS \subset I$ and so $I = (\bigcup_{x \varepsilon I} Sx \cup aS)$. Since S is connected so too is aS and Sx for each x, by (3.4). Also, $ax \varepsilon aS \cap Sx$ for each $x \varepsilon I$ and consequently, using result (1.16) of Chapter 1, I is connected.$\square$

The importance of the identity in the previous proposition is emphasized by pointing out what strong properties, which a connected semigroup S can possess, may fail to yield connectedness for ideals of S. The semigroup S = [0, 1/2] in the example preceding (3.4) is both topologically compact and algebraically abelian. (A semigroup S is <u>abelian</u> if and only if the equality ab = ba holds for all elements a and b in S.) Also, to illustrate that semigroups with these properties may have topologically very different disconnected ideals, let us mention another compact, connected, abelian semigroup: Consider the set S = {(x,y); $0 \leq x \leq 1$, $0 \leq y \leq 1$}, that is, a unit square in the plane (see Figure 3.2). Let S have the

usual topology and for any two elements (a,b) and (c,d) of S

define the multiplication * to be (a,b) * (c,d) = (0,bd). The

set I = {(x,y); x = 0, 1 and 0 $\leq$ y $\leq$ 1} is a disconnected closed

ideal of S and J = {(x,y); 0 $\leq$ x < 1/4, 3/4 < x $\leq$ 1 and 0 $\leq$ y $\leq$ 1}

is a disconnected open ideal (see Figure 3.2).

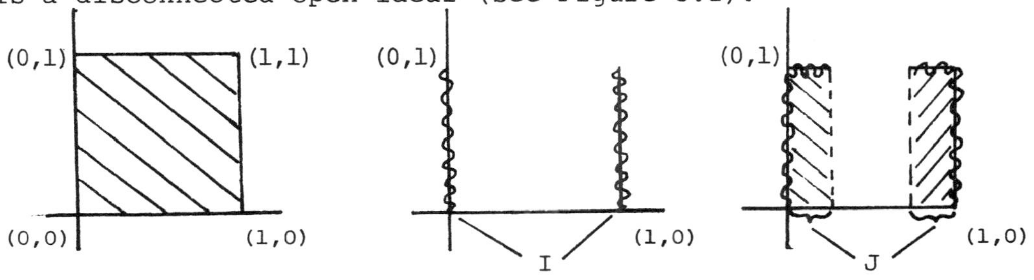

Figure 3.2

Even though a connected semigroup S may fail to have an

identity, nevertheless it is possible to say that some of its

ideals are connected:

3.6. <u>Proposition</u>. If I is an ideal of a connected semi-

group, then one and only one component of I is an ideal.

Proof. If x is any element of the ideal I, then, using

(3.4), it is easy to see that SxS is a connected subset of

I. Let C be the component of I containing SxS for a fixed

element x ε I. Both of the connected sets SC and CS intersect

SxS because they contain S(SxS) and (SxS)S, respectively. Thus,

since C is a component, (SC $\cup$ CS) $\subseteq$ C making C an ideal.

To show the uniqueness of such a component, let C' be

any component of I which is an ideal. The intersection

of the ideals SCS and SC'S is nonempty by (3.2). Thus,

since SCS is a subset of C, SC'S is connected and C is a
component, it follows that C contains SC'S, for otherwise
C would not be maximally connected. Therefore, $C \cap C' \neq \phi$
because SC'S is also a subset of the ideal C'. Since
components do not intersect by (1.21), one must conclude
that $C = C'$. $\square$

Section 3. Minimal Ideals

When optimality is considered in semigroups it is
somewhat strange that minimal ideals play a more exciting
role in the development of the theory than do maximal
ideals. For one thing we will see in this section that a
minimal ideal "inherits" very important topological properties from its "parent" semigroup, namely, compactness
and connectedness. In fact, in Chapter 4 we shall prove
that the compactness of a semigroup ensures the existence
of a (unique) minimal ideal and that additional properties
of a compact semigroup follow from the structure of its
minimal ideal. Much of this section establishes important
algebraic structure results concerning minimal ideals such
as the uniqueness of a minimal ideal, disjointness of
minimal left ideals and the fact that a minimal ideal is
the union of all the minimal left ideals provided the
semigroup has a minimal left ideal.

A left ideal of S is a <u>minimal left ideal</u> of S if it
does not properly contain any left ideals of S. Similarly,
an ideal of S is a <u>minimal ideal</u> (or <u>kernel</u>) of S if it

does not properly contain any ideals of S.

3.7. (algebraic) <u>Proposition</u>. If a semigroup contains a minimal ideal K, then K is unique.

Proof. Let K and K' be minimal ideals of S. KK' is an ideal and is contained in both K and K'. Thus KK' = K = K'. $\square$

3.8. (algebraic) <u>Proposition</u>. Distinct minimal left ideals are disjoint.

Proof. We will prove the contrapositive, namely, if two minimal left ideals are not disjoint, then they are not distinct. Let L_1 and L_2 be minimal left ideals such that $L_1 \cap L_2 \neq \phi$. The (b) part of (3.2) says $L_1 \cap L_2$ is a left ideal and so the fact that $L_1 \cap L_2$ is contained in L_1 implies $L_1 \cap L_2 = L_1$ by the minimality of L_1. Similarly, $L_1 \cap L_2 = L_2$ so that $L_1 = L_2$. $\square$

The next two results will be needed to establish (3.11) and (3.12).

3.9. (algebraic) <u>Proposition</u>. (a) If N is a minimal left ideal in a semigroup S, then N = Sa for any a ε N.

(b) If K is a kernel of S, then K = SaS for any a ε K.

Proof. (a) Sa $\subseteq$ SN $\subseteq$ N. The facts that Sa is a left ideal and N is minimal imply Sa = N. $\square$ The proof of part (b) is left as Exercise 3.23.

3.10. <u>Proposition</u>. If A and B are compact subsets of a semigroup, then AB is compact.

The proof of (3.10) parallels that of (3.4) and, there-

54

fore, is also included in Exercise 3.21.

We are now in a position to assert that a minimal ideal "inherits" the topological properties of compactness and connectedness from its "parent" semigroup. The proofs of the next three results are left as exercises. The existence of a minimal ideal is part of the hypothesis of the (3.12) theorem, but, in the case when the semigroup is compact, this premise will be seen, in Chapter 4, to be redundant. (An identical remark may be made about the hypothesis of (3.11).)

3.11. <u>Proposition</u>. If a semigroup is connected (compact), then a minimal left ideal, if one exists, is connected (compact).

3.12. <u>Theorem</u>. If a semigroup is connected (compact), then the kernel, if one exists, is connected (compact).

3.13. <u>Corollary</u>. If a semigroup is compact, then each minimal left ideal and the kernel are closed.

In the final propositions of this section our attention will now turn from the topological properties of minimal ideals to their algebraic structure. Most important of these results is the last theorem, (3.16), which says that a minimal ideal is the union of all the minimal left ideals provided the semigroup has at least one left minimal ideal. However, the presence of a single minimal left ideal yields an even more surprising result, namely, the presence of a minimal left ideal in every left ideal. This proposition ensuring a "global" conclusion from a "local" premise, as well as (3.16), requires a preliminary lemma:

3.14. (algebraic) <u>Lemma</u>. Suppose N is a minimal left ideal of a semigroup S. Then (a) Na is a minimal left ideal for any element a in S; and (b) every minimal left ideal may be expressed in the form Na for some a in S.

Proof. (a) Clearly, Na is a left ideal. If L is a left ideal contained in Na, then L = Ba where B = {x; x ε N and xa ε L}. SB $\subseteq$ SN $\subseteq$ N and (SB)a = S(Ba) = SL $\subseteq$ L so that SB $\subseteq$ B. Hence B is a left ideal of S contained in N and thus B = N by minimality. Therefore, L = Na.$\square$

The proof of part (b) of (3.14) will be an exercise at the end of this chapter.

3.15. (algebraic) <u>Proposition</u>. If a semigroup has a minimal left ideal, then every left ideal contains at least one minimal left ideal.

Proof. Let N be the existing minimal left ideal of the semigroup, consider any left ideal L and suppose b is any element of L. Then Nb $\subseteq$ NL $\subseteq$ L and, by the 3.14 Lemma, Nb is a minimal left ideal.$\square$

3.16. (algebraic) <u>Theorem</u>. If a semigroup S has a minimal left ideal, then S has a kernel, namely, the union of all minimal left ideals.

Proof. Let N be a minimal left ideal of S and consider NS, which is the union of all minimal left ideals by (3.14). Clearly, NS is an ideal. If I is an ideal and a is any element of S, then INa = (IN)a $\subseteq$ Na. Since INa is a left ideal, this implies that INa = Na because Na is a minimal left ideal. Therefore, Na = INa = I(Na) $\subseteq$ I. Since this is true for any a ε S we have NS $\subseteq$ I so that NS is contained in every ideal. We conclude that K = NS.$\square$

We may summarize the last two results graphically: Let S be a connected semigroup which has a kernel and, for the sake of simplicity, suppose that S has only two proper intersecting left ideals L_1 and L_2. Let N_1 and N_2 be the minimal left ideals of L_1 and L_2, respectively.

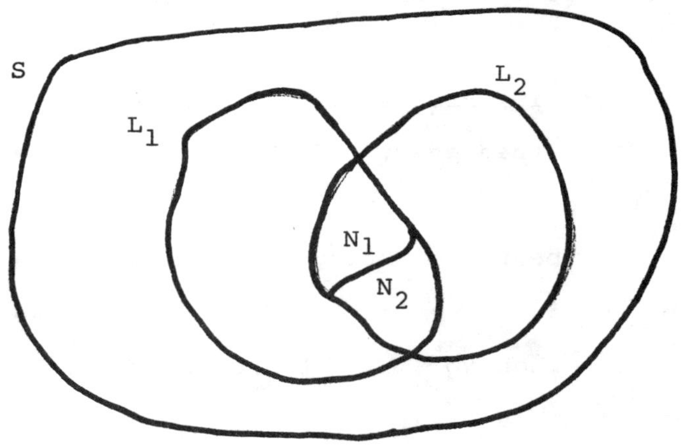

$$K = N_1 \cup N_2$$

Figure 3.3

EXERCISES

3.17. (algebraic) Prove the (d) part of (3.2).

3.18. Find examples of (a) a simple semigroup; (b)
an open proper ideal of the closed unit interval [0,1] (with
the relative topology); (c) a closed proper ideal which is
not a kernel; (d) a semigroup with no kernel; (e) a non-
compact semigroup which has a kernel; and (f) an open
minimal ideal.

3.19. (algebraic) Prove that a group cannot have any
proper left ideals.

3.20. (algebraic) Let I be an ideal of a semigroup S.
Prove that if I is a group, then I is both the kernel of S
and a maximal subgroup.

3.21. Prove Propositions (3.4) and (3.10), namely, if
A and B are connected (compact) subsets of a semigroup,
then AB is connected (compact).

3.22. (a) Verify that the subset D of the plane, discussed at the beginning of section 2, is a semigroup. (b)
Use part (a) of (3.2) to verify that $\{(1/2,\ 0),\ (-\ 1/2,\ 0)\}$
is a left ideal of D. (c) Why does our discussion of D not
violate the 3.5 Proposition?

3.23. Prove part (b) of (3.9), namely, if K is a kernel
of a semigroup S, then K = SaS for any a in K.

3.24. Prove the following results: (a) (3.11); (b)
(3.12); and (c) (3.13).

3.25. Let S be a compact semigroup with an identity e.

Prove that if e is in the kernel K, then S = K and, also,
S is a topological group. (Hint: To show that S is a
topological group one may verify and use the fact that K
is both a minimal left and a minimal right ideal.) When
viewing the position of elements of a semigroup, this exercise
becomes important. For, if e is the identity of a compact
semigroup which is not a group, then e cannot be in the
minimal ideal.

3.26. (algebraic) Prove part (b) of (3.14).

3.27. Let S be a connected semigroup containing left
ideals L_1 and L_2. Let N_1 and N_2 be minimal left ideals in
L_1 and L_2, respectively. What is wrong with each of the
following three diagrams?

(a)

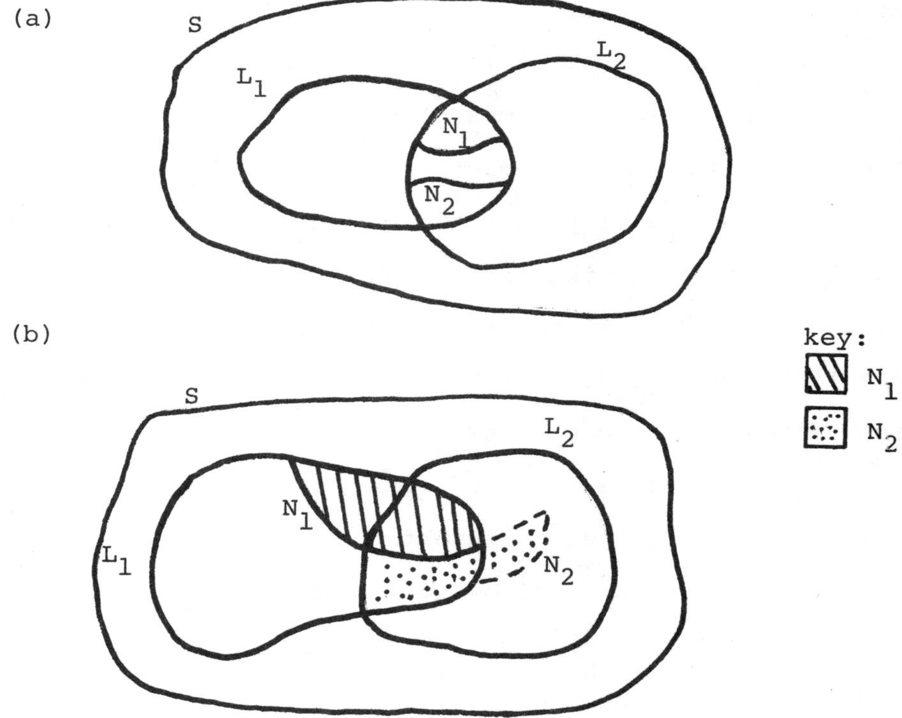

(c)   In this diagram let N_1 = A∪B and N_2 = B∪C.

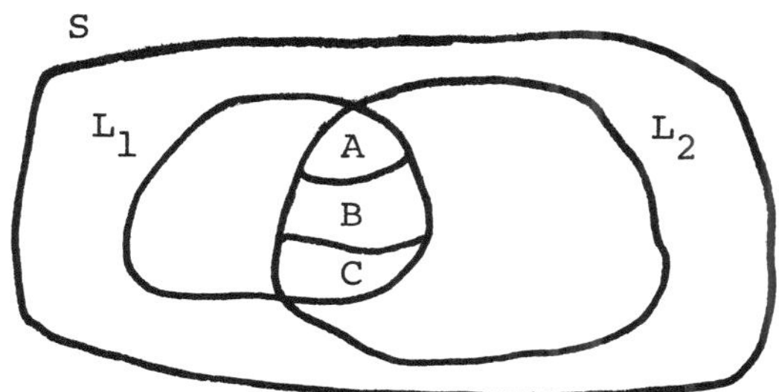

3.28.   (algebraic) Prove: If a semigroup S has a kernel K, then K is contained in every ideal of S.

3.29.   Determine, if possible, the kernel of the multiplicative interval semigroup (0, 1).

3.30.   (algebraic) Prove: A semigroup S is simple if and only if SaS = S for all a in S.

3.31.   (algebraic) Prove: An ideal contains every minimal left ideal.

CHAPTER IV. IDEALS IN A COMPACT SEMIGROUP

Section 1. Existence Theorems

In this section two important existence theorems resulting from topological compactness are derived. The first theorem, the existence of a kernel, is obtained chiefly by means of Zorn's Lemma. The second result, the presence of an idempotent, is derived in two very different ways; both methods contain significant ideas in their approaches.

4.1. <u>Theorem</u>. If S is a compact semigroup, then S contains at least one minimal left ideal.

Proof. Let L be any left ideal of S and let $\mathcal{C}$ be the collection of closed left ideals contained in L. $\mathcal{C}$ is nonempty because for any $b \in L$, Sb is a closed left ideal contained in L. We may partially order $\mathcal{C}$ by inclusion. Suppose that $\mathcal{T}$ is any tower (i.e., linearly ordered sub-collection) in $\mathcal{C}$. Clearly, $\mathcal{T}$ has the finite intersection property and so, since S is compact, $\cap\,\mathcal{T}$ is nonempty. (By $\cap\,\mathcal{T}$ we mean $\cap\,\{T;\ T \in \mathcal{T}\}$). Thus $\mathcal{T}$ has a lower bound in $\mathcal{C}$ so that $\mathcal{C}$ has a minimal element M by Zorn's Lemma. Since, by (3.13), a minimal left ideal of a compact semi-group is closed, we have that M is the desired minimal left ideal. (That M is a minimal left ideal may also be shown directly: If L_1 is a left ideal contained in M, let $x \in L_1$. Sx is a closed left ideal and $Sx \subset L_1 \subset M$ so that $Sx = L_1 = M$ by minimality of M. Thus M is a minimal left ideal in L.)$\square$

4.2. <u>Theorem</u>. Each compact semigroup S has a kernel.

The proof of (4.2) will be left as the (4.19) Exercise.

In Chapter 3 we saw that a kernel, if one exists, is the union of all the minimal left ideals. Here compactness ensures the existence of minimal left ideals and minimal right ideals. As a consequence, we can show that the kernel is the union of groups; but first we need two lemmas:

4.3. <u>Lemma</u>. If Γ is any index set, then $(\bigcup_{\alpha \in \Gamma} A_{\alpha})B = \bigcup_{\alpha \in \Gamma}(A_{\alpha} B)$.

Proof. $(\bigcup A_{\alpha}) B = m[(\bigcup A_{\alpha}) \times B] = m[\bigcup (A_{\alpha} \times B)] = \bigcup m(A_{\alpha} \times B) = \bigcup (A_{\alpha}B)$. (The next to the last equality holds because, if f is any function, we have the general result that $f(\bigcup D_{\alpha}) = \bigcup f(D_{\alpha})$. The proof of this set-theoretic result is trivial, by means of set inclusion, and the reader should verify it if he has not encountered it before.)$\square$

4.4. (algebraic) <u>Lemma</u>. If S has a minimal left ideal N and a minimal right ideal P, then $N \cap P$ is a group and, if e is the identity of this group, then $eSe = N \cap P$.

Proof. $PN \subset P \cap N$ so that $P \cap N \neq \phi$. Now define $\mathcal{R}$ to be the collection of all minimal right ideals and similarly $\mathcal{L}$ for minimal left ideals. Clearly, $\bigcup\{N \cap R; R \in \mathcal{R}\} \subset N$. Now consider $x \in N$. Since, by the 3.16 Theorem, $K = \bigcup\{L; L \in \mathcal{L}\} = \bigcup\{R; R \in \mathcal{R}\}$, we have x in some R. Therefore, we have equality, namely, $N = \bigcup\{N \cap R; R \in \mathcal{R}\}$.

Let $a \in N \cap P$. We have $(N \cap P)a \subset Sa \subset SN \subset N$ and $(N \cap P)a \subset Pa \subset PS \subset P$ so that $(N \cap P)a \subset N \cap P$. Now notice that $\bigcup\{N \cap R;$

62

$R \in \mathcal{R}\} = N = Na = [\cup\{N \cap R;\ R \in \mathcal{R}\}]a = \cup\{(N \cap R)a;\ R \in \mathcal{R}\}$.
We note that $(N \cap R_1) \cap (N \cap R_2) = \phi$ for any R_1, R_2 because
$R_1 \cap R_2 = \phi$. Therefore, since $(N \cap R)a \subset N \cap R$ for each R, in
order to have the above equality, namely, $\cup\{N \cap R; R \in \mathcal{R}\} = \cup\{(N \cap R)a;\ R \in \mathcal{R}\}$, it is necessary that $N \cap R \subset (N \cap R)a$
for each R. In particular, $N \cap P \subset (N \cap P)a$.

In an analogous fashion, $a(N \cap P) = N \cap P$. Therefore,
$N \cap P$ is a group.

Let e be the identity of $N \cap P$. By part (a) of (3.9)
and its right counterpart, $N \cap P = Se \cap eS \supset eSe$. Also, $N \cap P = e(N \cap P) \subset eN = e(Se) = eSe$. Therefore, $N \cap P = eSe$.$\square$

Let e be an idempotent. By $\underline{H[e]}$ we mean the maximal
subgroup containing e.

4.5. <u>Theorem</u>. If S is compact, then the kernel is a
union of groups, namely, $K = \cup\{H[e];\ e \in K \cap E\}$.

Proof. Compactness of S ensures the existence of both
minimal left and minimal right ideals. Thus $K = \cup\{R;\ R \in \mathcal{R}\} = \cup\{L;\ L \in \mathcal{L}\}$ and, as a result, $K = \cup\{L \cap R;\ L \in \mathcal{L},\ R \in \mathcal{R}\}$,
that is, by the lemma, the kernel is a union of groups. The
rest follows easily: For any $L \in \mathcal{L}$ and $R \in \mathcal{R}$, let e be
the identity of $L \cap R$. Clearly, $L \cap R \subset H[e]$. Also, $H[e] = eH[e]e \subset eSe = L \cap R$ and, consequently, $H[e] = R \cap L$.$\square$

It should be noted that (4.5) is the topological counterpart of the algebraic Rees-Suschkewitsch (R-S) structure theorem
for completely simple semigroups. (A semigroup S is <u>completely</u>
<u>simple</u> if S is simple and if S contains both a minimal left

ideal and a minimal right ideal; there is an alternative definition in terms of "primitive" idempotents.) In a formulation due to Alfred H. Clifford the R-S theorem states that if a semigroup S has both a minimal left ideal and a minimal right ideal, then S has a minimal ideal K which is the union of groups. The historical development is of interest here for it shows why the term kernel became a synonym for minimal ideal and gives the reason for our definition of a completely simple semigroup: Rees showed that the Suschkewitsch "kerngruppen," which was restricted to finite semigroups, is indeed the minimal ideal and that the class of all such kernels is identical with the class of finite simple semigroups without zero together with the zero semigroup of order 1. (It is remarked that Rees excludes the zero semigroup of order 2 from the class of simple semigroups.) In a later formulation given above, Clifford replaced the finiteness of the semigroup S by the condition that S contain at least one minimal left (or right) ideal and he showed that in such a semigroup the kernel K is completely simple in the sense of Rees, that is, according to the "primitive" idempotent definition.

One significant result of topological semigroups is contained in the 4.5 theorem, but it is not stated in its usual form: Every compact semigroup has an idempotent, namely, any identity of a subgroup in the kernel. However, I choose to present another, more direct, approach to this result for two reasons: (1) I wish to present the original

and traditional approach; and (2) the preliminary results
and methodology of this alternative approach are of interest.
Some readers may choose to omit this additional material
and should, if this is the case, proceed to the 4.12 Corollary.
We begin with a topological lemma which will be needed:

4.6. <u>Lemma</u>. If α is a function from $S \times S$ to S
defined by $\alpha(a,b) = ba$, then α is continuous.

Proof. Define $\tau:S \times S \to S \times S$ by $\tau(a,b) = (\pi_2(a,b),$
$\pi_1(a,b))$. By (2.5) τ is continuous and, clearly, $\tau(a,b) =$
(b,a). Now, note that $\alpha = m \circ \tau$ and so α is continuous.$\square$

4.7. <u>Proposition</u>. If A is an abelian subset of a
semigroup S [i.e., a, b ϵ A implies ab = ba], then A* is an
abelian subset of S.

Proof. Define a function $\theta:S \times S \to S \times S$ by $\theta(a,b) =$
(ab, ba). Each of the coordinates is a continuous function,
using (4.6), so that θ is continuous by the 2.5 lemma.
Clearly, $\theta(A \times A)$ is contained in the diagonal, Δ, of
$\theta(S \times S)$. (For brevity, let $\Delta(\theta) = \Delta$ of $\theta(S \times S)$.) We
have $[\theta(A \times A)]* \subset \Delta(\theta)$ because $\Delta(\theta)$ is closed. Since θ
is continuous, it follows that $\theta(A* \times A*) = \theta((A \times A)*) \subset$
$[\theta(A \times A)]*$, and so $\theta(A* \times A*) \subset \Delta(\theta)$ making A* abelian.$\square$

For any element x ϵ S, we define <u>0(x)</u> to be the set of
all positive powers of x, i.e., $0(x) = x, x^2, x^3, \ldots$.
Also, we define <u>$\Gamma(x)$</u> $= [0(x)]*$. Clearly, $0(x)$ is a semigroup and, by (2.4), so too is $\Gamma(x)$. In fact, it is easy
to show that $0(x)$ is the smallest semigroup containing x

and $\Gamma(x)$ is the smallest closed semigroup containing x.
The proof of the next proposition is left as the 4.22
Exercise.

4.8. <u>Proposition</u>. $\Gamma(x)$ is abelian.

4.9. (algebraic) <u>Proposition</u>. If an abelian semigroup
has a kernel K, then K is a group.

Proof. Let $x \in K$ and note that $xK \subset K$. Now $SxKS = x(SKS)$
because S is abelian and so $SxKS \subset xK$ so that xK is an ideal.
The minimality of K then implies that $xK = K$. Since S is
abelian, K is then a group by definition.☐

4.10. <u>Theorem</u>. A semigroup S has an idempotent if and
only if there exists an $x \in S$ such that $\Gamma(x)$ is compact.

Proof. Suppose S has an idempotent f. $0(f) = f$ so that
$\Gamma(f) = f$ which is clearly compact.

Now suppose that there exists an $x \in S$ such that $\Gamma(x)$
is compact. By (4.2) $\Gamma(x)$ has a kernel J (i.e., $\Gamma(x) J \Gamma(x)$
$\subset J$) and by the 4.8 Proposition $\Gamma(x)$ is abelian. Consequently,
by (4.9) J is a group and the identity of J is the desired
idempotent.☐

4.11. <u>Theorem</u>. Each compact semigroup has an idempotent.

In view of preceding results, the proof of (4.11) will
be left as an exercise at the end of this chapter.

4.12. <u>Corollary</u>. Every finite semigroup has an idempotent.

Proof. Every finite semigroup is compact.☐

Before ending this section we characterize algebraically
the closed semigroups of a compact group by showing that they

66

too are groups. A preliminary result is presented first:

4.13. _Lemma_. If S is a semigroup which is a group and
if C is a compact subsemigroup of S, then C is a group.

Proof. C compact implies that C has an idempotent f
by (4.11). We know that $f = 1$, the identity of S, because
a group has a unique idempotent. Now for any $x \in C$, xC is
compact and so contains 1. Thus $C = 1C \subseteq (xC)C = x(C^2) \subseteq xC$,
i.e., $C \subseteq xC$. Since C is a semigroup we have that $xC \subseteq C$ and
so the equality $C = xC$ holds. In a similar fashion $C = Cx$.$\square$

It is easy to find an example of a semigroup with an
identity which contains a compact subsemigroup which is not
a group. (4.28 Exercise.) As a consequence, the hypothesis
of the preceding lemma cannot be weakened substantially
(to S having only an identity).

4.14. _Proposition_. If a compact semigroup is a group,
then all of its closed subsemigroups are also groups.

Proof. This result is immediate in view of the lemma
since each closed subsemigroup is compact in this instance.$\square$

Section 2. Maximal Ideals.

A large portion of Chapter Three was devoted to the
significance of minimal ideals in semigroups. If one adheres
to the modern rule of "fair play" in all things, some mention
of maximal ideals is in order. However, it should be pointed
out that up to the present time maximal ideals have not played
as important a role in the development of the theory of semigroups as have minimal ideals. Nevertheless, there are

several interesting results concerning maximal ideals which are worthy of mention. Such a discussion is presented here, rather than in Chapter Three, because the main result, (4.16), is concerned with ideals in a compact semigroup.

If $A \subseteq S$, let $\underline{T(A)}$ be the union of all ideals of S contained in A. It is noted that if $T(A)$ is nonempty, then it is the largest ideal of S contained in A.

4.15. <u>Lemma</u>. If S is compact and A is an open subset of S, then $T(A)$ is open.

Proof. (The result is clearly true if $T(A)$ is empty.) Suppose $T(A) \neq \phi$, say $x \in T(A)$. Then $J(x) = \{x\} \cup xS \cup Sx \cup SxS$ is an ideal and it is contained in $T(A)$ because x is in the ideal $T(A)$. A is open implies there is an open set V_1 such that $x \in V_1 \subseteq A$. Since S is compact and A is open, Wallace's Theorem yields an open V_2 containing x such that $V_2 S \subseteq A$. Two more applications of Wallace's Theorem yield open sets V_3 and V_4 containing x such that $(V_1 \cup V_2 S \cup SV_3 \cup SV_4 S) \subseteq A$. Now let $V = V_1 \cap V_2 \cap V_3 \cap V_4$. Then V contains x and $(V \cup VS \cup SV \cup SVS) \subseteq A$. Therefore, since $x \in V \subseteq J(V) \subseteq T(A)$ it follows that $T(A)$ is open (by the 1.30 Exercise of Chapter 1). $\square$

It is noted that if A is closed, then $T(A)$ is closed. The brief proof of this fact is left as the (4.31) Exercise.

We define an ideal I to be a <u>maximal ideal</u> of S if the only ideals containing I are I and S. Also, for the sake of the next theorem, we mention a common topological term: A subset A of a space X is said to be <u>dense</u> in X if and only if

$A* = X$.

We are now equipped to state an existence theorem concerning maximal ideals. The proof of (4.16) will lean heavily on Zorn's Lemma, as did that of (4.1) which helped to establish the presence of kernels in compact semigroups.

4.16. <u>Theorem</u>. If a semigroup S is compact, then any proper ideal I of S is contained in a maximal proper ideal M, M is open and, further, M is either closed or it is dense in S.

Proof. First we shall prove that any maximal proper ideal M, if one exists, is open. We claim that $M = T(S\setminus\{x\})$ for any $x \notin M$: Since $T(S\setminus\{x\})$ is the union of all ideals contained in $S\setminus\{x\}$, it is evident that $M \subset T(S\setminus\{x\})$. Then, because $T(S\setminus\{x\})$ is a proper ideal containing M and M is a maximal proper ideal, it easily follows that $M = T(S\setminus\{x\})$. Since $S\setminus\{x\}$ is open and S is compact, the 4.15 Lemma implies that $T(S\setminus\{x\}) = M$ is open. Consequently, to see if a proper ideal I is contained in a maximal proper ideal, it suffices to consider only the open proper ideals containing I:

Let $x \in S\setminus I$. By (4.15) $T(S\setminus\{x\})$ is open and, clearly, it contains I. Let $\mathscr{A}$ be the collection of open proper ideals containing I and partially order $\mathscr{A}$ by inclusion. If $\mathscr{T}$ is a tower in $\mathscr{A}$, then $T_o = \cup\, \{T;\ T \in \mathscr{T}\}$ is an open ideal. If $S = T_o$, then $S = \bigcup_{j=1}^{n} T_j$ because S is compact so that $S = T_k$ for some k, a situation which cannot be because

$T_k \in \mathcal{A}$. We conclude that T_O is a proper open ideal and so is evidently an upper bound for $\mathcal{T}$ in $\mathcal{A}$. Therefore, by Zorn's Lemma, $\mathcal{A}$ has a maximal element, say M. Thus M is a maximal, (open) proper ideal containing I.

To finish the proof one notices that M* is an ideal by the 3.1 Proposition and, since $M \subseteq M^*$, it follows that the maximality of M as a proper ideal implies M = M* or M* = S.$\square$

The preceding discussion may be easily modified to show that the 4.16 Theorem also holds true for left ideals. That is, if S is compact, then any proper left ideal L of S is contained in a maximal proper left ideal L_M, L_M is open, and, in addition, L_M is either closed or it is dense in S.

So far our approach to maximal ideals has centered upon individual ideals in the semigroup rather than upon the semigroup itself. In a sense, our approach has been "local" in nature rather than "global". We now ask the question, does a semigroup have a proper ideal which contains every other proper ideal? The term maximal ideal would not be appropriate for such an ideal because we have already defined a maximal ideal to be something quite different. In the discussion to follow, if a semigroup has a proper ideal which contains every other proper ideal, we shall denote such an ideal by $\underline{M}_c$. First, we will present algebraic conditions which ensure that a semigroup has an M_c. Then, among other things, we point out, by example, that not even the presence of the topological properties of compactness and connectedness is enough to guarantee the existence of an M_c.

4.17. (algebraic) <u>Proposition</u>. If S is not simple and if S has at least one element q such that SqS = S, then S has an M_c.

Proof. Let Q = {q; SqS = S} and suppose S\Q = ϕ, that is, Q = S. Let A be any proper subset of S and note that SAS = S{a; SaS = S}S = $\underset{a \varepsilon A}{U}$ SaS = S so that A is not an ideal. Therefore S is simple contrary to hypothesis. We conclude that S\Q $\neq \phi$. Now suppose S\Q is not an ideal, i.e., S(S\Q)S $\cap$ Q $\neq \phi$. Then sas' ε Q for some a ε S\Q and s, s' ε S. As a result, (Ss)a(s'S) $\subset$ SaS $\subset$ S = S(sas')S and so a ε Q, a contradiction. Consequently, S\Q must be an ideal. Clearly, no subset of Q can be a proper ideal. Therefore, since (S\Q) $\cup$ q, for any q ε Q, is easily seen not to be a proper ideal, we have S\Q = M_c. $\square$

If a semigroup is compact and connected, then a result which is stronger than the converse of (4.17) may be stated, namely, if, in addition to these two topological properties S has an M_c, then S = SeS where e is an idempotent. This converse has limited structural significance and therefore, since its proof will require a number of preliminary results, we omit it and refer the interested reader to pages 45-50 of [2].

The necessity of the existence of an element q such that SqS = S in the 4.17 Proposition is shown by means of an example: Let S be the closed interval of real numbers [-1, 1] (with the discrete topology). Define a "multiplication"

* on S as follows: $x * y = xy$ if $x \geq 0$, $y \geq 0$

$x * y = 0$ if $x \leq 0$, $y \geq 0$ or if $x \geq 0$, $y \leq 0$

$x * y = -xy$ if $x \leq 0$, $y \leq 0$

where xy is the usual product of x and y. That * is surjective is emphasized in Figure 4.1.

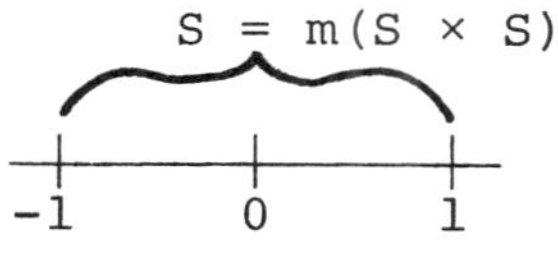

Figure 4.1

It is easy to verify that 1 is not an identity under this multiplication. The sets [-1, 1) and (-1, 1] are both maximal ideals, but M_c does not exist. If we assign the usual topology to this set, S will still be a topological semigroup. With this in mind, it is interesting to note that the strong conditions of compactness and connectedness are not sufficient to ensure the existence of an M_c.

EXERCISES

4.18. Prove: (a) If a semigroup S is the union of its left minimal ideals, then S is simple; (b) If S is compact, then the converse of (a) is true.

4.19. Prove (4.2).

4.20. Using the finite intersection property (Section 6, Chapter 1), prove that in a compact semigroup there is a unique minimal ideal and it is closed. (Hint: Consider the collection of closed ideals.)

4.21. In the proof of the 4.7 Proposition it is mentioned that $\Delta(\theta)$ is closed. Verify this fact.

4.22. Prove (4.8).

4.23. Verify that (i) $0(x)$ is a semigroup and the smallest one containing x; (ii) $\Gamma(x)$ is a semigroup and the smallest closed semigroup containing x.

4.24. Prove that if a compact semigroup is abelian, then its kernel is a topological group.

4.25. Prove that a compact, abelian, simple semigroup is a group.

4.26. In the proof of the 4.10 Proposition it is stated that $\Gamma(f) = f$. Why is this fact "obvious"?

4.27. Prove (4.11).

4.28. Find an example of a semigroup with an identity which has a compact subsemigroup which is not a group.

4.29. Verify that closed subsemigroups referred to in (4.14) are also topological groups.

4.30. (a) (algebraic) Prove that if S is a semigroup which is contained in a finite group G, then S is a subgroup of G. (b) Discuss possible relationships between part (a) and the 4.14 Proposition.

4.31. Let A be a topologically closed subset of a semigroup. Prove that $T(A)$ is closed.

4.32. Characterize the maximal proper ideals of the interval semigroup [0, 1/2] with the usual topology (and multiplication).

4.33. Find an example of a maximal proper closed ideal.

4.34. Find a semigroup which has no proper maximal ideals.

4.35. Verify that the 4.17 Proposition infers that the unit interval [0, 1], with the usual multiplication, has an M_c.

4.36. Consider the following erroneous statement: If S is a compact and connected semigroup, then S has a unique minimal left ideal, namely, the kernel K. (a) Find what is wrong with the following argument which tries to prove that statement: Each minimal left ideal and K are connected because S is connected and the fact that S is compact implies that each minimal left ideal is closed. Also, we know, algebraically, that distinct minimal left ideals are disjoint. As a result, any two minimal left ideals are disconnected. Therefore, the union of all minimal left ideals, namely, K, is disconnected--a contradiction--unless there is only one minimal

left ideal, say N. That is, K = N. (b) Find a counterexample to show that the statement is false.

CHAPTER V.   HOMOMORPHIC CONSTRUCTIONS

Section 1. Basic Homomorphism Results

The concepts of isomorphism and homomorphism are fundamental to algebraic structures and a semigroup is no exception.
Some of the more important semigroups involve this notion of
a homomorphism and so we begin by enumerating some of its
properties.

In the following let S be a semigroup with ordinary
multiplication and T be a semigroup (not necessarily distinct
from S) under an operation denoted by $*$. If a function
$h:S \to T$ has the property that, for any a, b in S, $h(ab) =$
$h(a)*h(b)$, then h is called a _homomorphism_. In the particular
case that $h(S) = T$, then such a homomorphism is often called
a _surmorphism_. If, in addition, h is a 1-1 function, then
it is an _isomorphism_. For a different, but related, definition, let $f:S \to T$ and let $A \subseteq S$ have algebraic structure--say
A is a subsemigroup, subgroup or ideal--then f is said to
respect that structure if $f(A)$ has the same structure in T
for all such A in S. For example, if for all ideals A of S
it turns out that $f(A)$ is also an ideal, then f is said to
respect ideals. One more concept will be needed for the
Homomorphism Theorem: If $f:S \to T$, we define $f^{-1}:T \to S$ by
$f^{-1}(a) = \{x \in S;\ f(x) = a\}$. Therefore, in general, one may
view f^{-1} as a multi-valued function, i.e., a relation. Notice
that the same symbol, say a^{-1}, is used to denote the multiplicative inverse of the element a in a group. The context

determines which meaning is correct.

5.1. (algebraic) <u>Lemma</u>. Let S and T be semigroups both with usual multiplication as the operation, and $h:S \to T$ be a homomorphism.

(i) If A_1, A_2 are subsets of S, then $h(A_1 A_2) = h(A_1)h(A_2)$;

(ii) If B_1, B_2 are subsets of T, then $h^{-1}(B_1 B_2) \supset h^{-1}(B_1)h^{-1}(B_2)$.

Proof. (i) Let $y \varepsilon h(A_1 A_2)$, say $y = h(xy)$ for $x \varepsilon A_1$, $y \varepsilon A_2$. Since h is a homomorphism, $y = h(x)h(y)$ and so $y \varepsilon h(A_1)h(A_2)$. The other set inclusion has an identical argument, using the homomorphism property "in reverse" and is omitted.

(ii) Consider the fact that $h(h^{-1}(B_1)h^{-1}(B_2)) = hh^{-1}(B_1)hh^{-1}(B_2) \subset B_1 B_2$. (The first equality holds by (i) letting $A_1 = h^{-1}(B_1)$ and $A_2 = h^{-1}(B_2)$.) Therefore, the result follows by viewing the first and last members of this relationship. $\square$

5.2. (algebraic) <u>Homomorphism Theorem</u>. Let S and T be semigroups and $h:S \to T$ be a homomorphism. Then,

(a) h and h^{-1} respect semigroups;

(b) h^{-1} respects ideals;

(c) h respects subgroups.

Proof. (a) Let A be a subsemigroup of S (possibly equaling S), i.e., $A^2 \subset A$. Using (i) of (5.1) we have $h(A)h(A) = h(A^2) \subset h(A)$ so that $h(A)$ is a semigroup of T.

The fact that h^{-1} respects subsemigroups follows in a similar fashion from (ii) of (5.1).

(b) Let L be a left ideal of T. Using (ii) of (5.1) we find that $Sh^{-1}(L) = h^{-1}(T)h^{-1}(L) \subseteq h^{-1}(TL) \subseteq h^{-1}(L)$. An analogous result holds for right ideals and so, since an ideal must be both a left ideal and a right ideal, h^{-1} respects ideals.

(c) Let G be a subgroup of S and suppose $y \in f(G)$, i.e., $y = f(x)$ for some $x \in G$. Then $f(x)f(G) = f(xG) = f(G)$ and, also, $f(G)f(x) = f(G)$ so that $f(G)$ is a subgroup of T. $\square$

It is easy to see that not all homomorphisms respect ideals. Let T be the multiplicative semigroup consisting of two idempotent elements 0 and 1 such that 0 is a zero. Let S be any semigroup and define $h:S \to T$ by $h(S) = 1$. Clearly, 1 is not an ideal of T. This same example shows that h^{-1} need not respect subgroups (if S is chosen, judiciously, so as not to be a group). Note that if T is changed to $\{1\}$, h^{-1} still does not respect subgroups; moreover, if S is then chosen to be a nontrivial semigroup with zero and T remains a singleton, then the fact that h^{-1} need not respect a minimal ideal is verified. This last sentence indicates that surmorphisms are not always helpful. However, the next theorem indicates that they can be very useful at times:

5.3. (algebraic) <u>Surmorphism Theorem</u>. Let S and T be semigroups and $p:S \to T$ be a surmorphism. Then,

78

(a)   p respects ideals;

(b)   p respects a minimal ideal (if one exists);

(c)   p respects an identity and a zero if one, or both,

exists in S.

Proof.   (a) Let I be an ideal of S.   Then, since $p(S) = T$, we have $Tp(I)T = p(S)p(I)p(S) = p(SIS) \subseteq p(I)$ which verifies that $p(I)$ is an ideal of T.

(b) Let K be the kernel of S. $p(K)$ is an ideal by the preceding paragraph.   Suppose J is an ideal contained in $p(K)$.   By (b) of (5.2) $p^{-1}(J)$ is an ideal.   Picking a $y \in J$ means $y = p(x)$ for some $x \in K$ and therefore, $x \in p^{-1}(J) \cap K$ making this latter set an ideal of S.   Consequently, $K \subseteq p^{-1}(J)$ by minimality of K and so $p(K) \subseteq p(p^{-1}(J)) = J$, the last equality holding because p is an "onto" function.   As a result $p(K)$ contains no proper ideals of T and so it is minimal.

(c) Let   e   be an identity for the semigroup S and let y be any element of T.   Since $p(S) = T$, $y = p(x)$ for some x in S.   Therefore, $p(e)y = p(e)p(x) = p(ex) = p(x) = y$. Similarly, $p(e)$ is a right identity.   If   0   is zero for S, the proof that $p(0)$ is a zero for T is entirely analogous to that just given for an identity.$\square$

We will end this section with a topological result concerning surmorphisms and compact spaces. $E(S)$ will denote the set of idempotents in S.

5.4.   <u>Proposition</u>.   Let S and T be semigroups.   If S

is compact and p:S → T is a continuous surmorphism, then
p(E(S)) = E(T).

Proof. It is easy to see that a homomorphism respects
idempotents (5.9 Exercise) and so p(E(S)) E(T). The rest
of the proof, namely, the other set inclusion, depends
primarily on the fact that a compact semigroup has an
idempotent: Let y ε E(T) and note that $p^{-1}(y)$ is nonempty
since p is a surmorphism. Topologically, {y} is closed
since T is Hausdorff and, therefore, $p^{-1}(y)$ is closed
because p is continuous. Also, $p^{-1}(y)$ is compact.
Algebraically, y, being an idempotent, is a subsemigroup
of T, and, by part (a) of (5.2), $p^{-1}(y)$ is a semigroup.
Now that we have established that $p^{-1}(y)$ is a compact semigroup, we conclude that it has an idempotent, say f. Then
y = p(f) and E(T) ⊂ p(E(S)).□

Section 2. Congruences and Quotient Semigroups

In Section 2 of Chapter 1 we purposely avoided a
definition of a relation, leaving it to the reader's
intuition because the notion of a Cartesian Product had not
yet been defined. It is to our advantage now to point out
that a <u>relation</u> on a set Y may be defined as any subset of
Y × Y. Consequently, if we speak of a closed relation $\mathcal{E}$
on a semigroup S, as we shall do, we mean that $\mathcal{E}$ is a
topologically closed subset of S × S. Also, a relation $\mathcal{E}$.
on a set Y becomes an <u>equivalence relation</u> on Y iff the
following three properties hold: (1) $\mathcal{E}$ is reflexive,

i.e., (y,y) is in $\mathcal{E}$ for all y in Y; (2) $\mathcal{E}$ is symmetric, i.e., if (a,b) is in $\mathcal{E}$, then (b,a) is in $\mathcal{E}$; (3) $\mathcal{E}$ is transitive, i.e., if (a,b) and (b,c) are in $\mathcal{E}$, then (a,c) is in $\mathcal{E}$.

Let S be a semigroup and $\mathcal{E}$ be an equivalence relation on S. If A is a subset of S, $[A]$ will denote all the elements in S which are $\mathcal{E}$-related to any element in A, that is, $[A]$ will be the union of the equivalence classes containing elements of A. Note that $[A] = \pi_1((S \times A) \cap \mathcal{E})$. The collection of all equivalence classes which S is divided into by $\mathcal{E}$ is denoted by $S/\mathcal{E}$. Defining $n:S \to S/\mathcal{E}$ by $n(x) = [x]$, it is easy to verify that n is a surjective function, that is, $n(S) = S/\mathcal{E}$. n is called the <u>natural</u>, or <u>canonical</u>, function.

As a topology for the set $S/\mathcal{E}$ we select the quotient topology (see Section 4 of Chapter 1) with regard to the natural function n, that is, Z is an open set in $S/\mathcal{E}$ iff $n^{-1}(Z)$ is open in S. Recall that the natural function n will be automatically continuous when $S/\mathcal{E}$ has this quotient topology.

First of all, we wish to establish under certain conditions, namely, when the semigroup S is compact and $\mathcal{E}$ is a closed equivalence relation, that $S/\mathcal{E}$ is a Hausdorff space. We begin with a lemma showing that the natural map n is closed under these conditions:

5.5. <u>Lemma</u>. If S is a compact semigroup and if $\mathcal{E}$

is a closed equivalence relation on S, then n is a closed
function.

Proof. Let A be a closed subset of S. Using results
(1.7), (1.8) and (1.10) of Chapter 1, S × S is compact and
(S × A) ∩ $\mathscr{E}$ is also compact. Since the projection π_1 is
continuous, [A] = π_1((S × A) ∩ $\mathscr{E}$) is compact and, therefore,
closed because S × S is Hausdorff by (1.5). In view of the
fact that [A] is the union of the equivalence classes containing elements of A, we see that n(A) = [A] and so n is
a closed function under the conditions set forth in this
lemma.□

5.6. <u>Proposition</u>. If S is a compact semigroup and
if $\mathscr{E}$ is a closed equivalence on S, then the quotient space
S/$\mathscr{E}$ is Hausdorff.

Proof. It is easy to see that each point, i.e.,
equivalence class, of S/$\mathscr{E}$ is closed because S is Hausdorff
and n is a closed function (by the preceding lemma). If
x and y are any distinct points of S/$\mathscr{E}$, then $n^{-1}(x)$ and
$n^{-1}(y)$ are closed because n is continuous. Since S is
normal by the 1.14 Corollary, there exist disjoint open
sets U and V containing $n^{-1}(x)$ and $n^{-1}(y)$, respectively.
Again using the fact that n is a closed function, we see
that n(S\U) and n(S\V) are closed subsets of S/$\mathscr{E}$. We
claim that W_1 = (S/$\mathscr{E}$)\n(S\U) and W_2 = (S/$\mathscr{E}$)\n(S\V) are
disjoint open sets containing x and y, respectively. It is
an easy matter to verify that x is in W_1 and y is in W_2,

and so we will prove here only the disjointness of W_1 and W_2: Beginning with one of DeMorgan's Laws for sets, we have $W_1 \cap W_2 = (S/\mathcal{E}) \setminus [n(S\setminus U) \cup n(S\setminus V)] = (S/\mathcal{E}) \setminus [n\{(S\setminus U) \cup (S\setminus V)\}] = (S/\mathcal{E}) \setminus n(S) = (S/\mathcal{E}) \setminus (S/\mathcal{E}) = \phi$, the next-to-last equality holding because n is surjective. Therefore, $S/\mathcal{E}$ is Hausdorff by definition. $\square$

So far we have only dealt with the topological aspects of $S/\mathcal{E}$. If the equivalence is also a congruence, a term to be defined in the next sentence, then $S/\mathcal{E}$ may be given an algebraic structure, and a semigroup, known as a quotient semigroup, will result under the conditions of (5.6). An equivalence relation $\mathcal{E}$ on a semigroup S is called a congruence if for each element (r,t) in $\mathcal{E}$ and every x in S we have that both (xr,xt) and (rx,tx) are elements of $\mathcal{E}$. This is often expressed by saying $(\Delta \mathcal{E} \cup \mathcal{E} \Delta) \subset \mathcal{E}$, where Δ is the diagonal of $S \times S$.

In order to show when $S/\mathcal{E}$ is a semigroup, the following topological lemma will be needed:

5.7. <u>Lemma</u>. If for each $\lambda \in \Lambda$, an indexing set, we are given a continuous function $g_\lambda : X_\lambda \to Y_\lambda$, then, letting $X = \times\{X_\lambda ; \lambda \in \Lambda\}$ and $Y = \times\{Y_\lambda ; \lambda \in \Lambda\}$, we may define a continuous function $g:X \to Y$ such that $\pi_\lambda g = g_\lambda \pi_\lambda$ for all $\lambda \in \Lambda$.

Proof. Results from previous chapters yield this lemma in an easy fashion. Let us keep the diagram of the conclusion in mind.

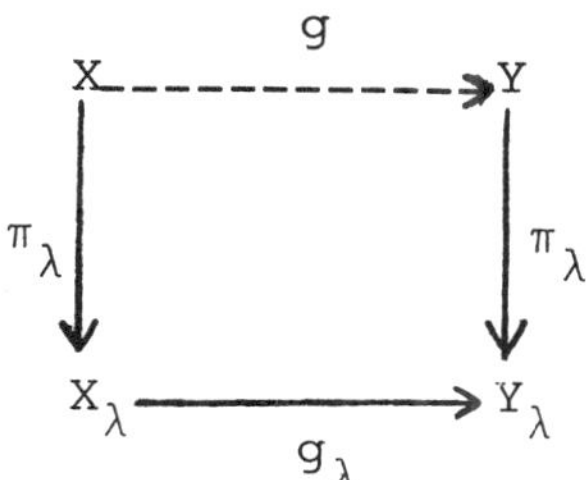

$g_\lambda \pi_\lambda : X \to Y_\lambda$ is continuous, for each λ, by (1.2) of Chapter 1. Then, by the 2.5 Lemma in Chapter 2, with $f_\lambda = g_\lambda \pi_\lambda$, there exists a continuous function g with the property $g_\lambda \pi_\lambda = \pi_\lambda g.\square$

If the index set Λ is a finite set of natural numbers, such as $\{1,2,3,....,k\}$, then the function g of (5.7) may be expressed as $g = (g_1, g_2, g_3, ..., g_k)$.

5.8. <u>Theorem</u>. If S is compact, $\mathscr{C}$ is a closed congruence on S and $n:S \to S/\mathscr{C}$ is the natural map, then we may define a multiplication μ on $S/\mathscr{C}$ in such a manner that n is a homomorphism and, therefore, $S/\mathscr{C}$ will be a semigroup.

Proof. We begin by defining a multiplication for $S/\mathscr{C}$. Consider the following diagram where m denotes the multiplication for S and $n \times n:S \times S \to S/\mathscr{C} \times S/\mathscr{C}$ is defined by $(n \times n)(a,b) = (n(a),n(b))$.

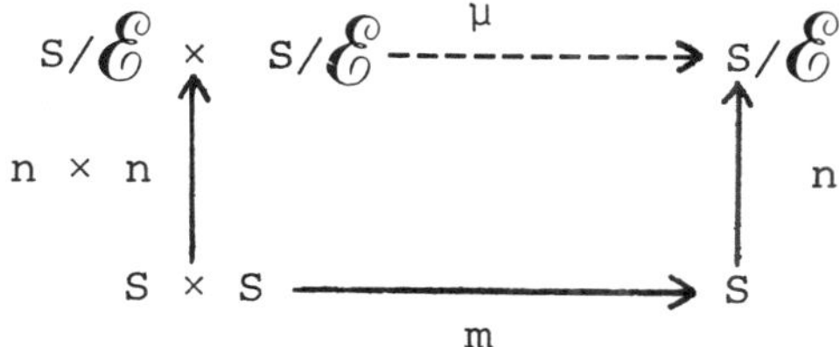

84

For A and B elements in $S/\mathcal{E}$ and a ϵ A, b ϵ B, we define
$\mu:S/\mathcal{E} \times S/\mathcal{E} \to S/\mathcal{E}$ by $\mu(A,B) = \mu(n(a),n(b)) = n(m(a,b))$.
μ "extends" the diagram as indicated by the dashed arrow.
Upon reflection, it is easy to see that this definition,
or one like it, is necessary if n is to be considered as
a homomorphism because $n(a)n(b) = AB = \mu(A,B)$ which by this
definition would become $n(m(a,b)) = n(ab)$. Consequently,
n will automatically be a homomorphism if we can show
that μ is "well-defined", that is, $\mu(A,B)$ is uniquely
determined. To verify this fact select any elements a,
a' in A and b, b' in B. We need to show that $n(ab) =$
$n(a'b')$ and it is at this point that one uses the congruence
property of $\mathcal{E}$:(a,a') in $\mathcal{E}$ implies that (ab,a'b) is in $\mathcal{E}$
and also (b,b') in $\mathcal{E}$ implies (a'b,a'b') in $\mathcal{E}$. Now,
since $\mathcal{E}$ is an equivalence relation (and, therefore, transitive), we have (ab,a'b') in $\mathcal{E}$ so that $n(ab) = n(a'b')$.

Consequently, in view of the Homomorphism Theorem
and (5.6), it suffices to have μ continuous for $S/\mathcal{E}$
to be a semigroup. We proceed to verify this continuity:
If Z is a closed subset of $S/\mathcal{E}$, then, using the continuity
of n, $n^{-1}(Z)$ is closed in S and, therefore, $m^{-1}n^{-1}(Z)$ is
closed since m too is continuous. Next we note that $n \times n$
is a closed map because $n \times n$ is continuous, by the 5.7
Lemma, and S is compact. Since $\mu^{-1}(Z) = (n \times n)[m^{-1}n^{-1}(Z)]$
(see 5.12 Exercise), we conclude that $\mu^{-1}(Z)$ is closed and
so μ is continuous. $\square$

The most common way to form a quotient semigroup is via an ideal I. (I × I) ∪ Δ constitutes an equivalence relation, on a semigroup S, in which I is an equivalence class and each point outside of I is a singleton equivalence class by itself. Because of the "engulfing" multiplication properties of an ideal, this relation is a congruence and an algebraic quotient semigroup exists. This semigroup is called the Rees quotient semigroup, in honor of the mathematician D. Rees, and it is denoted by S/I even though I is really not the congruence. For S/I to be a (topological) semigroup it is sufficient for I to be a closed ideal and S compact. It is easy to see that this Rees quotient S/I will have a zero, namely, the equivalence class I, whether or not S has a zero.

As an example, let S be the unit interval [0,1] with ordinary multiplication, select any element x in (0,1) and let I = [0,x]. Since the natural map n is continuous, S/I must be connected and so the effect of forming the quotient semigroup is to shrink [0,x] to the point x (figure 5.1). This shrinking effect is characteristic

Figure 5.1

of any Rees quotient S/I and thereby allows us to consider the properties of S independent of the engulfing multiplicative powers of that particular ideal I.

EXERCISES

5.9. (algebraic) Let S and S' be semigroups and h:S $\to$ S' be a homomorphism. Prove that h respects idempotents.

5.10. (algebraic) Show that if p:S $\to$ T is a surmorphism and S is simple, then T too is simple; in other words, establish that p respects simple semigroups.

5.11. In the proof of (5.6), verify that x is an element of W_1.

5.12. Both parts of this exercise deal with the (5.8) Theorem. (a) What result(s) concerning Hausdorff spaces are needed to show that n × n is a closed map? (b) Show that $\mu^{-1}(Z) = n \times n[m^{-1}n^{-1}(Z)]$.

5.13. Let S be a compact semigroup and I be a closed ideal of S. Prove that the congruence that I "induces" is closed.

5.14. (algebraic) Find a semigroup without a zero which has the property that one of its quotient semigroups has a zero.

CHAPTER VI.   GREEN'S EQUIVALENCE RELATIONS

Section 1.   The Quotient Space $S/\mathcal{H}$

One useful quotient space is formed by means of an equivalence known as Green's $\mathcal{H}$-relation; so-named in honor of the English mathematician James Alexander Green, who first studied this relation.   It will be beneficial to study the properties of the $\mathcal{H}$-relation, and some other equivalence relations, and so we begin by defining, for any subset A of a semigroup S, $L(A) = A \cup SA$ and $R(A) = A \cup AS$.   The relationship of these concepts to ideals is immediate:

6.1.   (algebraic) <u>Proposition</u>.   L(A) is the smallest left ideal containing A.

Proof. $S(A \cup SA) = SA \cup S^2 A = SA \subseteq L(A)$ so that L(A) is a left ideal.   Now let N be a left ideal containing A and contained in L(A).   Since $SN \subseteq N$, we have $SA \subseteq N$.   Therefore, $(A \cup SA) \subseteq N$ and, consequently, L(A) contains no proper left ideals which have A as a subset. $\square$

If we now let the A, of L(A), be a singleton subset, we are in a position to define the relation $\mathcal{H}$ as $\{(x,y) \ \varepsilon \ S \times S; \ L(x) = L(y) \text{ and } R(x) = R(y)\}$.   It is easy to verify that $\mathcal{H}$ is an equivalence relation.   For notation, let $H_a = \{x \ \varepsilon \ S; \ L(x) = L(a) \text{ and } R(x) = R(a)\}$ and refer to H_a as the <u>H-slice</u> through a. H_a constitutes one of the disjoint sets in the partition of S according to $\mathcal{H}$, namely, the one containing a, and so it is often also called an <u>H-class</u>.

For an example of the H-classes of a semigroup the reader is referred to pages 54-56 of [3], the text by Clifford and Preston. The example given there is the set of all functions of a set X into itself. The operation of semigroup is function composition, i.e., $gf(x) = g(f(x))$ and this semigroup is called the full transformation semigroup on X (because these functions are often referred to as transformations). Specifically, pages 54-56 detail a complete listing of the 71 H-classes of the full transformation semigroup on a set consisting of four elements. This semigroup contains $4^4 = 256$ functions as elements. (One may wish to view first pages 1 and 2 of [3] for a brief background discussion of transformations.)

For another example involving H-slices, consider the set consisting of the following nine points in the plane:

$$\overline{0} = (0,0) \ , \quad \overline{1} = (1,1) \ , \quad \overline{2} = (1,-1) \ ,$$
$$\overline{3} = (-1,1), \quad \overline{4} = (-1,-1), \quad \overline{5} = (1,0) \ ,$$
$$\overline{6} = (-1,0), \quad \overline{7} = (0,1) \ , \quad \overline{8} = (0,-1) \ .$$

This set becomes an interesting semigroup, call it N, if we consider coordinate-wise multiplication, i.e., $(a,b)(c,d) = (ac,bd)$. Its multiplication table is given next. It has a zero, an identity and it is abelian. Also, the elements $\overline{1}, \overline{2}, \overline{3}$ and $\overline{4}$ form a subgroup.

	$\overline{0}$	$\overline{1}$	$\overline{2}$	$\overline{3}$	$\overline{4}$	$\overline{5}$	$\overline{6}$	$\overline{7}$	$\overline{8}$
$\overline{0}$	$\overline{0}$	$\overline{0}$	$\overline{0}$	$\overline{0}$	$\overline{0}$	$\overline{0}$	$\overline{0}$	$\overline{0}$	$\overline{0}$
$\overline{1}$	$\overline{0}$	$\overline{1}$	$\overline{2}$	$\overline{3}$	$\overline{4}$	$\overline{5}$	$\overline{6}$	$\overline{7}$	$\overline{8}$
$\overline{2}$	$\overline{0}$	$\overline{2}$	$\overline{1}$	$\overline{4}$	$\overline{3}$	$\overline{5}$	$\overline{6}$	$\overline{8}$	$\overline{7}$
$\overline{3}$	$\overline{0}$	$\overline{3}$	$\overline{4}$	$\overline{1}$	$\overline{2}$	$\overline{6}$	$\overline{5}$	$\overline{7}$	$\overline{8}$
$\overline{4}$	$\overline{0}$	$\overline{4}$	$\overline{3}$	$\overline{2}$	$\overline{1}$	$\overline{6}$	$\overline{5}$	$\overline{8}$	$\overline{7}$
$\overline{5}$	$\overline{0}$	$\overline{5}$	$\overline{5}$	$\overline{6}$	$\overline{6}$	$\overline{5}$	$\overline{6}$	$\overline{0}$	$\overline{0}$
$\overline{6}$	$\overline{0}$	$\overline{6}$	$\overline{6}$	$\overline{5}$	$\overline{5}$	$\overline{6}$	$\overline{5}$	$\overline{0}$	$\overline{0}$
$\overline{7}$	$\overline{0}$	$\overline{7}$	$\overline{8}$	$\overline{7}$	$\overline{8}$	$\overline{0}$	$\overline{0}$	$\overline{7}$	$\overline{8}$
$\overline{8}$	$\overline{0}$	$\overline{8}$	$\overline{7}$	$\overline{8}$	$\overline{7}$	$\overline{0}$	$\overline{0}$	$\overline{8}$	$\overline{7}$

Since N is abelian, $L(x) = R(x)$ and $H_a = \{x \in S;\ L(x) = L(a)\}$. From the multiplication table it is easy to verify that $L(x) = S$ if and only if $x \in \{\overline{1},\overline{2},\overline{3},\overline{4}\}$. Therefore, this subgroup is an H-slice. The other three H classes are also easily verified to be: $\{\overline{0}\}$, $\{\overline{5},\overline{6}\}$, and $\{\overline{7},\overline{8}\}$. Moreover, in this case $\mathcal{H}$ is a congruence and so we can form the quotient semigroup $N/\mathcal{H}$. To prove directly that $\mathcal{H}$ is a congruence is easy but laborious. Let us just give a few details to indicate the procedure: Consider any two elements in an equivalence class, say $\overline{3}$ and $\overline{4}$. For any x in N, are $x\overline{3}$ and $x\overline{4}$ in the same H-slice? For instance, if $x = \overline{7}$, are $\overline{7}\cdot\overline{3} = \overline{7}$ and $\overline{7}\cdot\overline{4} = \overline{8}$ $\mathcal{H}$ related? The answer is, of course, yes. Following this procedure, there are many computations which are necessary to verify that $\mathcal{H}$ is a congruence here. However, later in Section 2 and Exercise (6.14), we will see that checking these computations is not necessary, in this case, because $\mathcal{H}$ is a congruence in view of the fact that N is abelian.

The quotient semigroup, N/$\mathcal{H}$, has the multiplication table given below where $\bar{\bar{0}} = \{\bar{0}\}$, $\bar{\bar{1}} = \{\bar{1},\bar{2},\bar{3},\bar{4}\}$, c = $\{\bar{5},\bar{6}\}$ and d = $\{\bar{7},\bar{8}\}$. Notice that the shrunken subgroup is the identity, $\bar{\bar{1}}$. Also, note that each element of this quotient semigroup is an idempotent.

	$\bar{\bar{0}}$	$\bar{\bar{1}}$	c	d
$\bar{\bar{0}}$	$\bar{\bar{0}}$	$\bar{\bar{0}}$	$\bar{\bar{0}}$	$\bar{\bar{0}}$
$\bar{\bar{1}}$	$\bar{\bar{0}}$	$\bar{\bar{1}}$	c	d
c	$\bar{\bar{0}}$	c	c	$\bar{\bar{0}}$
d	$\bar{\bar{0}}$	d	$\bar{\bar{0}}$	d

In order to form a quotient space using the $\mathcal{H}$-equivalence relation it will be necessary to have $\mathcal{H}$ closed for the quotient space to be Hausdorff. It is the object of the rest of this initial section to show that the compactness of S is sufficient for $\mathcal{H}$ to be closed.

Before stating the next proposition, several preparatory definitions and a topological lemma will be introduced. For any subsets A and B of a semigroup S, define $A^{(-1)}B = \{x \; \varepsilon \; S; \; Ax \cap B \neq \phi\}$ and $A^{[-1]}B = \{x \; \varepsilon \; S; \; Ax \subset B\}$. Note the equality of these two sets if A is a singleton subset.

6.2. <u>Lemma</u>. Let A and B be subsets of a semigroup.
(a) If A is compact and B is open, then $A^{[-1]}B$ is open.

(b) If B is open, then $A^{(-1)}B$ is open.

Proof. (a) Let x be an element of $A^{[-1]}B$. Since A and {x} are compact, we may use the first corollary to the Wallace Theorem, namely, (1.13) of Chapter 1, to obtain open sets U and V with the properties $A \subset U$, $x \in V$ and $UV \subset B$. Therefore, $AV \subset B$, that is, $V \subset A^{[-1]}B$ and so, by the (1.30) Exercise, $A^{[-1]}B$ is open. (b) Note that $A^{(-1)}B = \cup\{a^{(-1)}B;\ a \in A\}$ and that, by part (a), $a^{(-1)}B$ is open for each $a \in A$. The result is now immediate. $\square$

For the next proposition we define a relation $\mathscr{L}_0$ on a semigroup S: $\mathscr{L}_0 = \{(x,y);\ x,y \in S$ and $L(x) \subset L(y)\}$.

6.3. <u>Proposition</u>. If a semigroup S is compact, then $\mathscr{L}_0$ is closed.

Proof. Although relatively lengthy, this proof is not difficult. We will show that the complement of $\mathscr{L}_0$, namely, $(S \times S)\setminus \mathscr{L}_0$, is open. To begin, suppose that (x,y) is not in $\mathscr{L}_0$. Since this means that L(x) is not a subset of L(y), there is an element q in $L(x)\setminus L(y)$. L(y) is closed because {y} is compact and S is both compact and Hausdorff. Therefore, since S is normal, there are disjoint open sets V and V' such that $q \in V$ and $L(y) \subset V'$. Consequently, $V \cap L(x) \neq \phi$ so that $x \in V$ or $V \cap Sx$ is nonempty. This last fact may be expressed as $x \in V \cup S^{(-1)}V$. Letting $U = V \cup S^{(-1)}V$, we first notice that $S^{(-1)}V$ is open by part (b) of the preceding lemma and, therefore, deduce that U is open. Now the set containment $L(y) \subset V'$ implies that $y \in V' \cap S^{[-1]}V'$. Letting $U' = V' \cap S^{[-1]}V'$ and using part (a) of (6.2), it is easy to see that U' is open.

Of course, $(x,y) \in U \times U'$ and we will show that $U \times U' \subset (S \times S) \setminus \mathcal{L}_o$: Suppose the contrary, namely, there is a point (a,b) in $(U \times U') \cap \mathcal{L}_o$. Pause for a moment to reflect that $b \in U'$ implies $L(b) \subset V'$ and, therefore, $L(a) \subset V'$ because $(a,b) \in \mathcal{L}_o$. Now $a \in U$ means $a \in V$ or $Sa \cap V \neq \phi$. It is easy to verify that both of these possibilities contradict the fact that V and V' are disjoint sets. Therefore, one must conclude that $U \times U'$ and $\mathcal{L}_o$ do not intersect. It follows, by the (1.30) Exercise, that $(S \times S) \setminus \mathcal{L}_o$ is open and so $\mathcal{L}_o$ is closed. $\square$

Of course, $\mathcal{L}_o$ is not even an equivalence relation. We take a step closer to our goal by defining an equivalence allied to $\mathcal{L}_o$: $\mathcal{L} = \{(x,y) \in S \times S; \ L(x) = L(y)\}$. The next two results will be needed to show that $\mathcal{L}$ is closed if S is compact. The first just explicitly states a part of the (4.6) Lemma and the second gives a different formulation of $\mathcal{L}$.

6.4. <u>Lemma</u>. Let S be a semigroup. If $\tau : S \times S \to S \times S$ is defined by $\tau(x,y) = (y,x)$, then τ is continuous.

Proof. Kindly refer to the proof of (4.6).

6.5. (algebraic) <u>Lemma</u>. $\mathcal{L} = \mathcal{L}_o \cap \tau \mathcal{L}_o$.

Proof. Let $(x,y) \in \tau(\mathcal{L}_o)$. This means (y,x) is in $\mathcal{L}_o$, that is, $L(y) \subset L(x)$. Consequently, $\tau \mathcal{L}_o = \{(x,y) \in S \times S; \ L(y) \subset L(x)\}$. The result now becomes immediate. $\square$

6.6. <u>Proposition</u>. If S is a compact semigroup, then $\mathcal{L}$ is a closed equivalence relation on S.

The proof of (6.6) is left to the reader as Exercise (6.19).

In analogy to $\mathcal{L}$, we define $\mathcal{R} = \{(x,y) \in S \times S; \ R(x) =$

R(y)}. Now that all three have been defined, it should be mentioned that $\mathscr{L}$, $\mathscr{R}$ and $\mathscr{H}$ are <u>Green's Relations</u>. Please notice that $\mathscr{H} = \mathscr{L} \cap \mathscr{R}$. It is evident that analogous results, to those for $\mathscr{L}$, hold for $\mathscr{R}$. Therefore, the concluding theorem of this section is:

6.7. <u>Theorem</u>. If a semigroup S is compact, then $\mathscr{H}$ is a closed equivalence on S.

Proof. In view of (6.6), and its corresponding result for $\mathscr{R}$, this proof becomes trivial. $\square$

It is possible to generalize Green's Relations. Let T be any subset of S and define L(A,T) = A $\cup$ TA and $\mathscr{L}$(T) = {(x,y); L(x,T) = L(y,T)}. If T = T* (i.e., T is closed), then most of the previous results hold, namely, when S is compact one finds $\mathscr{L}$(T), $\mathscr{R}$(T) and $\mathscr{H}$(T) closed and so quotient spaces, which are Hausdorff, may be formed. $\mathscr{L}$(T), $\mathscr{R}$(T) and $\mathscr{H}$(T) are called <u>Relative Green Relations</u>. The connection between Green's Relations and this generalization goes beyond the fact that T may be S; for, relative Green Relations provide algebraic "decompositions" of the (usual) Green Relations in the following way: If we select a particular T in S, then each equivalence $\mathscr{L}$-class is the union of a number of $\mathscr{L}$(T)-classes. In this sense, the relative Green Relations enable us to do a "finer" analysis because the semigroup is (usually) partitioned into more disjoint subsets than in the case of $\mathscr{L}$, $\mathscr{R}$ and $\mathscr{H}$. In fact, if we first partition S according to one of Green's Relations, say $\mathscr{H}$, and next, for a particular subset T, form $\mathcal{H}(T)$, then we may say truly that we have formed a partition of a partition. Relative Green Relations have been studied extensively, mainly by A.D. Wallace [13] and [14].

Section 2. Uses of S/$\mathcal{H}$.

We are now in a position to consider the $\mathcal{H}$- relation as a possible candidate for generating quotient semigroups. Unfortunately, although $\mathcal{L}$ is a right congruence, that is, for all a in S we have (xa, ya) ε $\mathcal{L}$ whenever (x,y) ε $\mathcal{L}$, and $\mathcal{R}$ is a left congruence, nevertheless $\mathcal{H}$ need not be a congruence. Yet, in certain instances it will have this important property. In particular, when S is abelian $\mathcal{H}$ is a congruence; please notice that in this case R(x) = L(x) and the Green equivalence relations coincide, i.e., $\mathcal{H} = \mathcal{L} = \mathcal{R}$. In general though, as in our main theorem, (6.8), it is necessary to hypothesize the congruity of $\mathcal{H}$.

6.8. <u>Theorem</u>. If a semigroup S is compact and $\mathcal{H}$ is a congruence on S, then S/$\mathcal{H}$ is a semigroup.

Proof. This result follows directly from the (6.7) and (5.8) Theorems. $\square$

The quotient space, and sometimes semigroup, S/$\mathcal{H}$ is of special interest because all the subgroups of S are "shrunk" to single elements of the quotient space so that, in particular, when S/$\mathcal{H}$ is a semigroup we find that all its connected subgroups are degenerate, i.e., each has cardinality 1. (We choose not to prove these facts rigorously here. They are a consequence of (6.9) which follows soon.)

Whether or not $\mathcal{H}$ is a congruence, it does lead to a related quotient semigroup, and in fact a quotient group, under certain non-restrictive conditions. This occurs for an H-slice, say H_x, when S is compact, H_x is not a singleton set and $H_x^{[-1]}H_x$ is not empty. Consider the relation $\mathcal{S}(x) = \{(u,v);\ u,\ v\ \varepsilon\ H_x^{[-1]}H_x$ and $xu = xv\}$. It turns out that $\mathcal{S}(x)$ is a congruence and $H_x^{[-1]}H_x/\mathcal{S}(x)$ is a group, known as the Schutzenberger group (in honor of the French mathematician Marcel P. Schutzenberger). Verification of these remarks, though not very difficult, would take too long at this point and if the reader wishes, at a future time, he may consult [3], [9] and [14] for details.

The quotient space $S/\mathcal{H}$, while of interest in itself, also gives us a valuable technique to derive information about the semigroup from which it was constructed. We illustrate this statement, and conclude this section by showing that if S is compact, then the union of all its subgroups is closed. (It is noted that (2.11) is not sufficient to establish this fact unless we know that the number of subgroups is finite--which we don't.)

First we give an algebraic result stating that the H-slice H_e, for any idempotent e, is a maximal subgroup. For notation, let H[e] be the maximal subgroup of S containing e. It is easy to see that H[e] $= \{x\ \varepsilon\ eSe;\ xx' = x'x = e$ for some $x'\ \varepsilon\ eSe\}$.

6.9. (algebraic) <u>Proposition</u>. Let e be an idempotent of a semigroup. Then $H[e] = H_e$.

Proof. Suppose $x \in H[e]$ and denote its inverse by x^{-1}. Then, $x \cup Sx = xe \cup Sxe \subseteq Se \subseteq e \cup Se$. In a similar fashion, $R(x) \subseteq R(e)$. Also, $e \cup Se = x^{-1}x \cup Sx^{-1}x \subseteq Sx \subseteq x \cup Sx$ and, similarly, $R(e) \subseteq R(x)$. That is, $L(x) = L(e)$ and $R(x) = R(e)$ so that $x \in H_e$. Consequently, $H[e] \subseteq H_e$.

Now let $a \in H_e$ and $a \neq e$. Then, since $R(a) = R(e)$ and $L(a) = L(e)$, we have $a \in eS \cap Se$ implying that $a = eb = ce$ for some $b, c \in S$. Therefore, $ae = (ce)e = ce = a$ and likewise $ea = e(eb) = eb = a$. In addition, $e \in aS \cap Sa$ so that $e = ay = za$ for some $y, z \in S$. Next we find $e = e^2 = e(za) = (ez)(ea) = (eze)a$ and, similarly, $e = a(eye)$. However, it turns out that $eye = eze$ because $eye = e^2ye = (ezea)(eye) = (eze)(aeye) = eze^2 = eze$, and so $eye = a^{-1}$. Therefore, $a \in H[e]$ and $H_e \subseteq H[e]$ completing the proof. $\square$

This result says that the $\mathcal{H}$-equivalence relation partitions S in such a manner that each maximal subgroup is an equivalence class. Semigroups which are not groups are not as fortunate; for, according to a result known as Green's Theorem, if a and b are in H_c, then $ab \in H_c$ if and only if H_c is a subgroup. Therefore, semigroups which are not groups have their elements "scattered" throughout the $\mathcal{H}$-equivalence classes. (Since we will not use this information contained in Green's Theorem, we will omit its proof. Consult [3] for the details, if desired.)

Now we are ready to show that the union $\mathcal{G}$ of all subgroups of a compact semigroup is topologically closed:

6.10. <u>Theorem</u>. If S is a compact semigroup and $\mathcal{G}$ denotes the union of all subgroups of S, then $\mathcal{G}$ is closed.

Proof. Let n:S → S/$\mathcal{H}$ be the natural map, that is, the image of each point of S is the H-slice containing it. Every subgroup G contains an idempotent, namely, its identity, and so G is contained in a maximal subgroup H[e] which happens to be an H-slice by (6.9). Therefore, we make the assertion, which the reader may verify at his leisure, that $\mathcal{G} = n^{-1}(n(E))$ where E is the set of idempotents of S. Recall that E is closed by (2.6). Consequently, a simple sequence of steps, using the knowledge that S is both compact and Hausdorff and n is continuous, yields our conclusion.□

Let us mention that it is not necessary to use the idea of a quotient space to prove this theorem; for, if one notices that $\mathcal{G} = \cup \{H_e;\ e\ \varepsilon\ E\} = \pi_1(\mathcal{H} \cap S \times E)$, where π_1 is the first projection map of S × S onto S, then the result follows easily using (6.7). However, in view of our previous results, the proof we gave may be easier to grasp (and, perhaps, easier to find on one's own initiative). It is this last parenthetical remark which gives the signif-icance of the quotient space S/$\mathcal{H}$ in regards information

98

concerning S: It may not be a necessary tool, but it can
be a helpful and enlightening one.

EXERCISES

6.11. (algebraic) Let A be a subset of a semigroup S.
Show that, if $A^{[-1]}A$ is not empty, then it is a subsemigroup
of S.

6.12. Let X and Y be subsets of a semigroup S. Prove
that $X^{[-1]}Y$ is closed whenever Y is closed. (Hints: First
verify that $X^{[-1]}Y = \cap \{a^{(-1)}Y;\ a \in X\}$ and then notice that
the function $la:S \to S$, defined by $la(s) = as$, is continuous.)

6.13. In the proof of (6.3), the result showing $\mathcal{L}_o$
is closed when S is compact, fill in the details that verify
that L(y) is closed.

6.14. (algebraic) Show that $\mathcal{L}$ is a right congruence.
(An equivalence relation $\mathcal{E}$ on a semigroup S is a right
congruence iff for each element (r,t) in $\mathcal{E}$ and every x
in S we find that (rx,tx) is in $\mathcal{E}$.)

6.15. (algebraic) For any $T \subset S$, prove that the Green
Relation $\mathcal{L}$ is the union of $\mathcal{L}$(T)-classes.

6.16. (algebraic) Prove: If G is a subgroup of a
semigroup S, then G is an equivalence class in the relative Green
Relation $\mathcal{H}$(G) decomposition of S.

6.17. Let S be a compact semigroup. Prove the following
statements: (a) Each H-slice containing an idempotent is
closed. (b) The union of all H-slices containing idempotents

is closed.

6.18. (a) Defining $L_a = \{x; (x,a) \in \mathscr{L}\}$, show that L_a is closed if S is compact. (Hint: Letting π_1 and π_2 be the projections of $S \times S$ onto S, verify that $L_a = \pi_1(\pi_2^{-1}(a) \cap \mathscr{L})$.) (b) Prove that any H-slice, say H_a, is closed when S is compact. (This result is, of course, a generalization of (6.17a).)

6.19. Prove (6.6).

CHAPTER VII. SEMIGROUP CONSTRUCTIONS

Section 1. Semigroup products

To this point we have limited ourselves to classifying abstract semigroups and our examples have mostly dealt with the real number line. Because topology is concerned with the position of elements, it is natural to consider semigroup structures for subsets of R^2, the plane, and R^3, ordinary 3-dimensional space. Fortunately, continuous multiplications for such subsets can easily be attained by means of our previous results. The first construction we shall consider is the "semigroup product" of two semigroups and afterward examples relating to the plane and 3-dimensional space will be given.

The <u>semigroup product</u> of the two semigroups S and S' is the Cartesian product set S × S' together with a multiplication defined coordinatewise, that is, for a, b ε S and a', b' ε S' we have $(a,a')(b,b') = (ab,a'b')$. Alternatively, the semigroup product may be expressed as a function P:S × S' → S × S' which is associative and continuous. The associativity is immediate from that of S and S' since each coordinate of the product is concerned with a single semigroup; for instance, in our definition the first coordinate depends upon S and the second upon S'. The fact that m and m', the multiplications of S and S', respectively, are continuous easily infers the continuity of P = (m,m') by the topological

lemma (5.7).

Examples. (a) Using ordinary arithmetic multipli-

cation and the notation I = [0,1] and Q = [-1,1] it is now

possible to construct semigroups, by means of a semigroup

product, whose underlying sets in the plane are the squares

I × I and Q × Q.

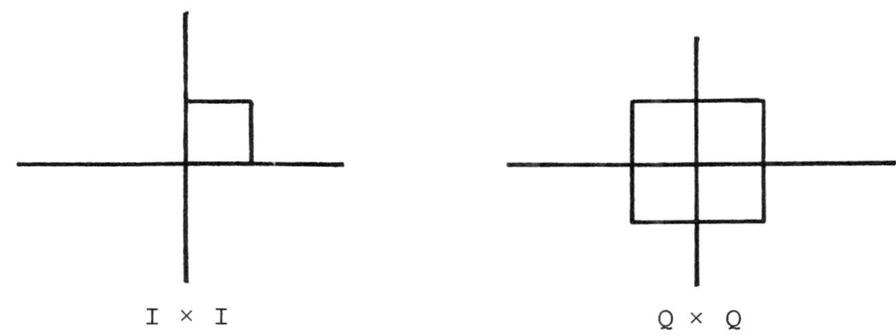

Figure 7.1(a)

(b) Consider in the plane the unit disk D_1 = $\{(x,y); \sqrt{x^2 + y^2} \leq 1\}$. ($D_1$ is often called a <u>2-cell</u>.) D_1 is a subset of Q × Q so that in order to show it is a semigroup it suffices, because of (1.25), to show that algebraically $(D_1)^2 \subset D_1$. Let (x,y) and (x',y') ε D_1. Since xx' ≤ x and yy' ≤ y, we have $\sqrt{(xx')^2 + (yy')^2} \leq \sqrt{x^2 + y^2} \leq 1$.

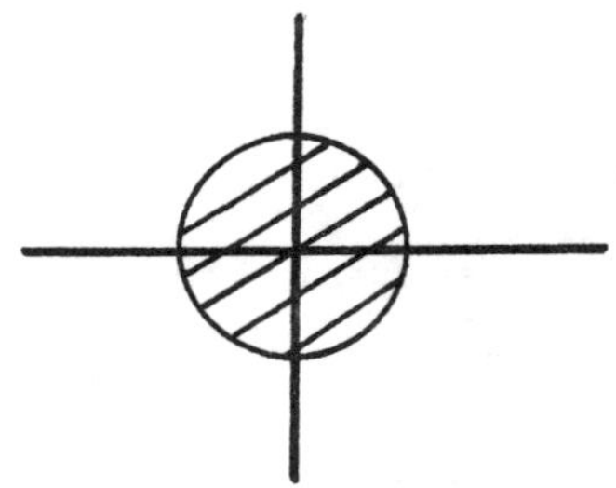

Figure 7.1(b)   The closed disk D_1

(c)   Now, letting $S = D_1$ and $S' = I$, we may use the semigroup product to define a semigroup structure for a 3-dimensional solid cylinder, $D_1 \times I$.

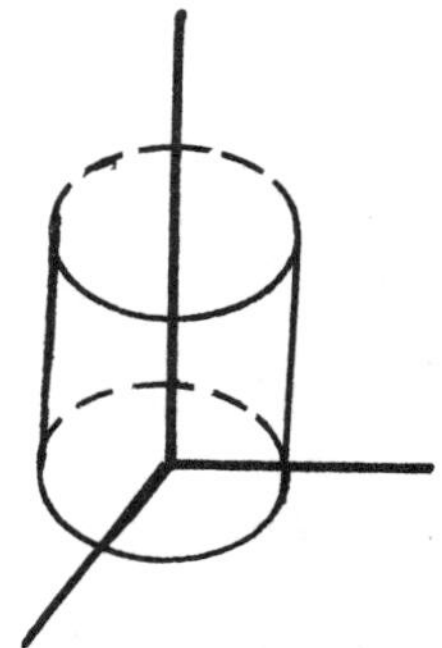

Figure 7.1(c) $D_1 \times I$

We may extend this notion of a semigroup product to more than two semigroups using (5.7). In particular, let $S = S' = S'' = I$ and in this manner define a semigroup structure for the 3-dimensional unit cube $I \times I \times I$.

Section 2. Identity and zero

In (2.1) we saw that in a semigroup the identity and zero, if they exist, are both unique. The real intervals $[0,1)$ and $(0,1]$, semigroups under multiplication, indicate, respectively, that neither an identity nor zero are necessary. Of course, $R(a)$ for any $a > 1$, or $(0,1)$, illustrates that both may be missing. In this section we will show that an identity, or zero, can easily be appended to semigroups which lack them.

Let S be any semigroup. Define the semigroup S^1 as follows: If S has an identity, $S^1 = S$; if not, $S^1 = S \cup \{1\}$, where $1 \notin S$, the multiplication is S remains unchanged and 1 acts as an identity for $S \cup \{1\}$. Let m and ϕ be the multiplications for S and S^1, respectively. To show the continuity of ϕ, one must explicitly state the topology $\mathscr{I}$ of S^1. A set V is open, i.e., $V \in \mathscr{I}$, iff V is in $\mathscr{S}$, the topology for S, or $V \setminus \{1\}$ is in $\mathscr{S}$. Note that $\{1\}$ is both an open and closed subset of S^1. Consider any closed set A contained in $\phi(S^1 \times S^1)$. If $1 \in A$, letting $A = B \cup 1$ and noting that $B \subset S$, we have $\phi^{-1}(A) = m^{-1}(B) \cup (1 \times B) \cup (B \times 1) \cup (1 \times 1)$; if $1 \notin A$, $\phi^{-1}(A)$ has the same form with $A = B$ except the singleton set $\{1 \times 1\}$ is not present. It is not difficult to verify that B is a closed subset of S and S^1 if $A = B \cup 1$ (and the proof is left for the reader to supply). As a result, by (1.3), the sets $(1 \times B)$, $(B \times 1)$, (1×1) and, consequently, $\phi^{-1}(A)$ are closed when 1 is in A. If 1 is not in A, it is easy to verify that A is a closed set of S and so, again using (1.3), $\phi^{-1}(A)$ is closed in this instance also. Therefore, by (1.1), ϕ is continuous.

A semigroup S^O may be defined similarly: Let S be any semigroup. If S has a zero and the cardinality of $S > 1$, $S^O = S$; otherwise, define the set $S^O = S \cup \{0\}$ and a multiplication function $\theta : S^O \times S^O \to S^O$ by $\theta(x,y) = m(x,y)$ if $x,y \in S$ and $\theta(x,y) = 0$ if $x = 0$ or $y = 0$. The reader should verify that, in a manner analogous to that for ϕ, θ too is continuous.

A word about some confusing terminology: If S is not a singleton set, if it has a zero and if the multiplication has the form $m(x,y) = 0$ for all x, $y \in S$, then such a semigroup is called a <u>zero semigroup</u>. Therefore, one should be careful NOT to call S^O a zero semigroup unless $m(x,y) = 0$ for all x, $y \in S$, an unlikely, coincidental occurrence.

If a semigroup has a zero, then a number of results become trivial because $\{0\}$ is the minimal ideal. For this reason, a study of simple semigroups which have a zero element is pointless and so the concept of a 0-simple semigroup emerges: A semigroup S is <u>0-simple</u> iff $S \neq \{0\}$ and $SaS = S$ for all nonzero elements a in S. (Compare this definition with the equivalent form of simplicity given in (3.29). It is easy to prove that S is 0-simple if and only if $S^2 \neq \{0\}$ and $\{0\}$ is the only proper ideal. Using this fact as an alternative definition it is easy to see that a semigroup S, with zero, which has {0} as its only proper ideal is either 0-simple or it is a zero semigroup of order 2, i.e., it consists of two elements.

Section 3. Borsuk's paste job and complex number semigroups

One particularly interesting construction technique, named in honor of the Polish mathematician Karol Borsuk, involves the ideas of Chapter 5. Begin with a compact semigroup S, a closed subsemigroup A of S, and a continuous surmorphism $p:A \to T$. (Recall that T too is a semigroup by (5.2).) The set $\mathscr{P} = \Delta(S) \cup \{(a,a') \text{ in } A \times A;\ p(a) = p(a')\}$ turns out to be a closed equivalence relation on S and so, by (5.8), $S/\mathscr{P}$ is a quotient semigroup when $\mathscr{P}$ is a congruence; moreover, $S/\mathscr{P}$ looks like S with A cut out and T pasted in as p directs. The congruity of $\mathscr{P}$ is assured when A is an ideal and both $p(xa) = p(xa')$ and $p(ax) = p(a'x)$ for all x in S and all (a,a') in $\mathscr{P}$. Verification of these well-known facts will not be presented here except for the fact that $\mathscr{P}$ is closed:

The subtle technique employed here is "functional" in nature. Define the function $p \times p:A \times A \to T \times T$ by $(p \times p)(a,a') = (p(a),p(a'))$, a definition entirely similar to that used in (5.8). $p \times p$ is surjective and, by (5.7), continuous. The diagonal of $T \times T$ is elements $(p(a),p(a'))$ with the property $p(a) = p(a')$ and so $(p \times p)^{-1}(\Delta) = \{(a,a') \text{ in } A \times A;\ p(a) = p(a')\}$. Since $\Delta(T \times T)$ is closed and p × p is continuous, the fact that $\mathscr{P}$ is closed now easily follows.

An example shows that the paste job may always be applied when the compact semigroup possesses a proper closed ideal A. Let T be the trivial group consisting of a single point and p:A → T. In this case $\mathscr{P}$ = Δ(S) ∪ (A × A) making the verification that $\mathscr{P}$ is closed an easy matter and the sufficient conditions listed above for congruity, namely, p(xa) = p(xa') and p(ax) = p(a'x), automatically hold.

In order to further illustrate this construction technique, we digress for a moment and discuss semigroup structures in the complex number plane. Since our intended audience may not have been exposed yet to a treatment of complex numbers $\mathbb{C}$, we briefly mention that we are talking about all ordered pairs of real numbers (a,b) with the operations of coordinatewise addition and <u>complex multi-plication</u> defined in terms of ordinary real number addition and multiplication by (a,b)(x,y) = (ax - by, ay + bx). Using the continuity of real addition and multiplication, the continuity of these operations follows from (2.5). With the convention i^2 = -1, the notation a + bi is often used to represent the complex number (a,b) because in this manner complex multiplication "acts like" real multiplication; for example, (a + bi)(x + yi) = ax - by + i(ay + bx). Of particular importance to us is the real number $\sqrt{x^2 + y^2}$

associated with the complex number $z = (x,y)$. This real number, known as the _modulus_ of z and written as $|z|$, may be visualized as the distance from the origin $(0,0)$ to the point (x,y). The fact that the modulus of the (complex) product of two complex numbers is the (real) product of their moduli, i.e., $|zz'| = |z| \, |z'|$, is of great significance for our purposes because it allows us to define a surmorphism from the complex plane to the real number line.

Several subsets of $\not{C}$ which form semigroups under multiplication are worthy of special mention: The unit disk $D_1' = \{z; \; |z| \leq 1\}$ (see Figure 7.2) and, for $r < 1$, $D_r' = \{z; \; |z| \leq r\}$. D_r' is a closed ideal of D_1'. Pause for a moment and consider similar subsets of the real Euclidean plane, namely, D_1, as defined in Section 1, and $D_r = \{(x,y); \sqrt{x^2 + y^2} \leq r\}$ for $r < 1$. Under the semigroup product, D_r too is a semigroup; in fact, it is a closed ideal of D_1. The reason why the complex plane is introduced is because the function $d: D_1 \to [0,1]$ defined by $d(x,y) = \sqrt{x^2 + y^2}$ is not a homomorphism as the reader may easily verify.

Now let $Y = \{z; \; z = (x,y) \text{ with } 0 \leq y < 2\pi \text{ radians}\}$. Using the result known as _Euler's relation_, $e^{iy} = \cos y + i(\sin y)$, one obtains as the image of the complex function $f(z) = e^{iy}$, with domain Y, a circle C' in the complex plane. Again, view Figure 7.2. C' is not only a semigroup, but is also a group known as the _circle group_.

(The corresponding circle in the Euclidean plane, C = $\{(x,y); \sqrt{x^2 + y^2} = 1\}$, known as the <u>one-sphere</u>, is not even a semigroup under the semigroup product operation.)

Figure 7.2

We are now in a position to continue our illustrations of Borsuk's paste job. Let S be D_1' and A be the complex disk $D_{1/2}'$, noting that A is a closed ideal of D_1'. Let T be the real interval [0, 1/2] and define p:A → T by p(z) = |z|. The semigroup obtained using the paste job may be visualized as a flat umbrella whose handle A/$\mathcal{C}$, which is a collection of equivalence classes, is a homeo-morphic copy of the interval [0, 1/2]. This situation is depicted in Figure 7.3 and there u and $\overline{o}$ represent the multiplicative identity (1,0) and zero (0,0), respectively.

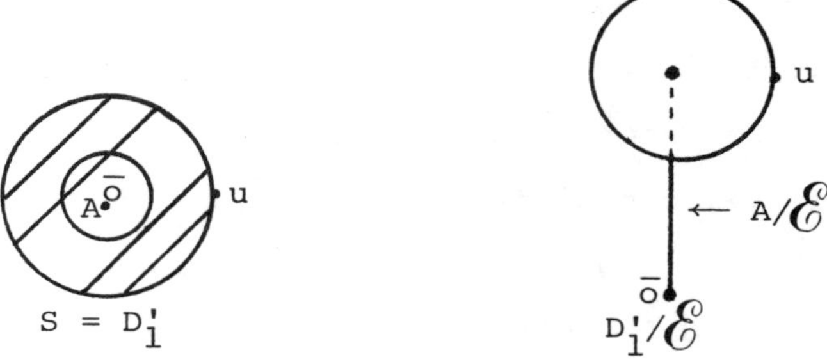

Figure 7.3

We conclude this section with the fleeting remark that this pasting procedure is very useful for constructing semigroups on strange spaces. No substantiation for this claim will be given here.

Section 4. Miscellaneous constructions

(a) By means of a Rees quotient it is possible to "convert" a disconnected semigroup S into a connected one if there is a closed ideal intersecting all the components of S. For example, letting F be a finite semigroup with n   elements and I = [0,1], F × I is a compact semigroup, using the semigroup product, and F × {0} is a closed ideal intersecting each of the   n   components of F × I. The quotient semigroup (F × I)/(F × {0}) is connected and may be viewed as the prongs of a hand fan.

Figure 7.4(a) Hand fan with n = 2.

(b) Now that complex multiplication has been introduced, we remark that C' × I, where I denotes the real interval [0,1] and C' the circle group, is a semigroup under the semigroup product multiplication and visually it may be pictured as a hollow cylinder, i.e., only the points on the surface. This cylindrical shell, C' × I, may be turned into a conical

shell which will also have a semigroup structure. For any

r < 1, the set T_r = C' × [0,r] is a closed ideal of C' × I

and the Rees quotient semigroup, which may then be formed,

must look like a cone because it is the continuous image

of the cylinder which is connected. Consequently, when

looking down from above the cone, (C' × I)/T_r is seen as

a (planar) 2-cell and, indeed, it turns out to be homeo-

morphic to D_1.

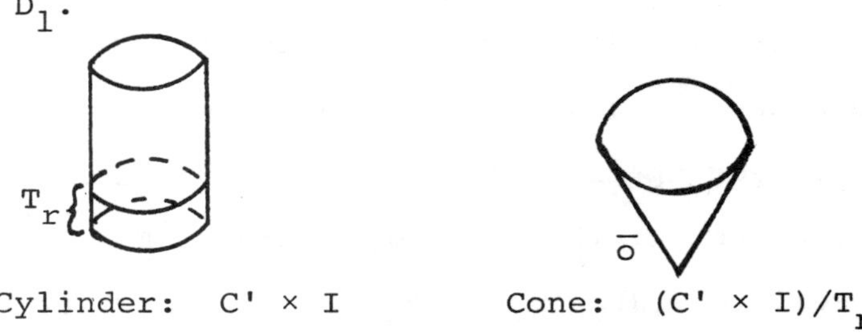

Cylinder: C' × I   Cone: (C' × I)/T_r

Figure 7.4(b) Hollow Cylindrical and Conical Semigroups

(c) In the (2.14) Exercise the notion of a rectangular

band semigroup was introduced. Now we will show that its

multiplication $\oplus$ is continuous. As in (2.14) let X and

Y be the sets under consideration. Let X_1 = X_2 = X × Y,

g_1 = π_1 and g_2 = π_2. Then by (5.7) $g((x_1,y_1), (x_2,y_2))$ =

$(\pi_1(x_1,y_1), \pi_2(x_2,y_2))$ = (x_1,y_2) is continuous.

(d) One of the fundamental ideas of topological semi-

groups, known as the Rees-Suschkewitsch-Wallace Theorem,

gives the topological structure of a kernel of a compact

semigroup. To begin, let us establish a necessary preliminary

result, namely, the formation of a particular semigroup.
Let S be any semigroup and define Z to be the cartesian
product $S \times S \times S$. We will verify that Z with the multiplication $(a,b,c)(x,y,z) = (a,bcxy,z)$ is a semigroup. To
show the continuity of the multiplication we again resort
to (5.7): Let $X_1 = X_3 = S$, $X_2 = S \times S \times S \times S$, g_1 and g_3
be identity maps and g_2 be an obvious extension of the
multiplication of S, namely, $g_2(a,b,c,d) = abcd$. Once
we show that g_2 is continuous our result will follow
immediately by (5.7) with the multiplication of Z being
expressed as $g = (g_1,g_2,g_3)$. For the continuity of g_2,
let m be the multiplication of S and define a function
$h:S \times S \times S \times S \to S \times S$ by $h(a,b,c,d) = (m(a,b),m(c,d)) =$
(ab,cd). h is continuous, by (5.7) with $X_1 = X_2 = S \times S$,
and so the composite function $mh = g_2$ is continuous by
(1.2).

Up until this time we have not bothered to check the
associativity of the multiplication functions of our newly
constructed semigroups either because it followed directly
from the underlying semigroup, as in the case of complex
multiplication; or because it followed in a trivial manner
as a consequence of the way in which the multiplication
was defined, for example, the rectangular band. However,
the peculiar form of Z necessitates a verification of this
associative property. Letting u,v,w be elements of Z, one
writes each in coordinate form; i.e., $u = (u_1,u_2,u_3)$ etc.,

and it then becomes a simple matter to verify that the coordinate forms of (uv)w and u(vw) are the same.

Although different in appearance, (4.5) and the Rees-Suschkewitsch-Wallace (R-S-W) Theorem are both structure theorems for compact, completely simple semigroups. In their conclusions, both results characterize the kernel K of such a semigroup. In order to state the (R-S-W) Theorem in its entirety we need to first introduce a topological concept, that of a "retract": A subset A, of a space X, is a retract iff there exists a continuous surjective function r:X → A with the property that the restriction of r to A is the identity. (Equivalently, A is a retract of X iff the identity map i:A → A can be "extended", i.e., its domain enlarged, in a continuous fashion to all of X.) As examples, the unit interval [0,1] is a retract of the set R of real numbers (as a consequence of an important result known as Tietze's Extension Theorem), the unit square I × I is a retract of R^2 and the unit cube I × I × I is a retract of R^3. However, in the plane the unit circle $\{(x,y); \sqrt{x^2 + y^2} = 1\}$ is not a retract of the disk $D_1 = \{(x,y); \sqrt{x^2 + y^2} \leq 1\}$.

Also, we define an isomorphism to be a function which is both a (topological) homeomorphism and an (algebraic) isomorphism. Then, the (R-S-W) Theorem states, for any idempotent e in the kernel K of a compact semigroup S, that K is isomorphic with the product $(Se \cap E) \times eSe \times (eS \cap E)$, where E represents the set of idempotents of S, with multiplication in the product defined by $(x,y,z)(x',y',z') = (x,yzx'y',z')$. In addition, the kernel K, the set $K \cap E$ and each of the "factors" of the product, namely, $(Se \cap E)$, eSe and $(eS \cap E)$, are all retracts of S. For brevity, we choose to omit its proof here. The advanced, interested reader may wish to consult [5] and [12] for some of the details.

(e) Significant is the fact that any topological Hausdorff space can be turned into a semigroup and, furthermore, this may be done in such an easy fashion that the multiplication is referred to often as the trivial multiplication. Define $xy = x$ for all x,y in the set W. The continuity is immediate since this is a projection function. (One may alternatively choose the second projection, that is, define $xy = y$.) At this point let us mention that it is known that if a space S is homeomorphic to the unit circle and it is a semigroup such that its multiplication gives $S^2 = S$, then this multiplication must have one of three forms: (a) $xy = x$, (b) $xy = y$ or (c) $S = C'$, the circle group discussed in Section 7.3.

In passing it is noted that the ability to introduce a semigroup structure on any set sharply contrasts with the topological group situation.

A constant function will also yield a semigroup in any Hausdorff space by selecting any element c in the space W

114

and defining xy = c for all x,y in W. (The continuity can
be established in a manner similar to (1.32).) However, it
matters where c is positioned, that is, different semigroups arise for various selections of c. For instance,
if the underlying space of S is a closed interval in R, it
matters whether c is an endpoint or an interior point.

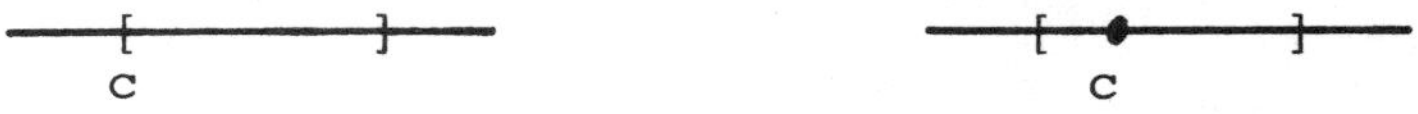

Nonisomorphic semigroups with multiplication xy = c

Figure 7.5

The main difference between the two multiplications,
xy = x and xy = c, is that xy = c is an abelian operation
and the other is not. However, neither of these semigroups
has an identity and one common question asked of topological
structures is, does an abelian, associative multiplication
with identity exist? (This type of structure is abbreviated
to CAMU where U stands for unit, a sometimes synonym for
identity.) Unlike the question of the existence of a
semigroup, which we have answered in the affirmative for
all spaces, establishing the existence of a CAMU for a
specific space often proves to be a difficult task.

Speaking of an identity brings to mind that topological
placement of algebraic entities is an important topic in
the study of semigroups. An identity, in general, turns
up as a boundary point of the set. Consider [0,1] under

ordinary multiplication or the complex semigroup D_1'. For another example, take a circle in the complex plane and attach to it a closed interval (see Figure 7.6). In order for this set to have a semigroup structure with an identity it is necessary, as can be shown by solely topological methods, that the identity must be the endpoint of the interval which is not attached to the circle. Again, see Figure 7.6. (Moreover, it can be proved algebraically that the circle must be a subgroup, namely, the circle group C'.)

Figure 7.6

(f) Another semigroup related to constant multiplication $xy = c$ and also to the translation function, mentioned in (6.12), is one using the multiplication $q(x,y) = xcy$ for a fixed c in the set W and for all x,y in W. Establishing the continuity of this multiplication q is of mild interest for its methodology. Consider the function $t:S \times S \to S \times S$ defined by $t(x,y) = (x,cy)$. t is continuous by (5.7), $t = (g_1,g_2)$, with $X_1 = X_2 = S$, g_1 the identity map and g_2 the "left translation" function lc. (lc is continuous by (1.25) because it is a restriction of the multiplication, m, of S to the set $\{c\} \times S$.) Now we turn to a function $r:S \times cS \to S$ defined by $r(x,cy) = xcy$.

r is continuous because it is a restriction of m and so
the composite map rt, which is q, is also continuous by
(1.2).

(g) The existence of an entity known as a semi-
topological group was mentioned in Chapter 2. To review,
a semi-topological group is a (topological) semigroup
which is algebraically a group, but not a topological
group. Examples have been postponed until now in order
to keep Chapter 2 as simple as possible. They are considered
here because they illustrate one of the basic reasons for
creating new semigroups, namely, determining that specific
structures are possible.

For the first example we view the set of real numbers,
R, under ordinary addition. For the topology of R we select
not the usual topology, but instead one whose base $\mathcal{B}$ is
{[a,b); a,b in R}. This system is both a (topological)
semigroup and an additive algebraic group. However, the
inversion function i(x) = -x is not continuous because
the inverse image, under i, of the open set [a,b) is
(-b,-a] which is not open.

A second example is given in [8] and deals with the
intersection of all L^p-spaces based on the unit interval
[0,1] for $p \geq 1$.

(h) Most of the results of this text apply to semigroups which are compact because most of the research has
dealt with this class of semigroups. However, many important

semigroups are not compact, for example, the set R of real numbers with the usual topology and the interval (0,1) with the relative topology, both semigroups under ordinary multiplication. As one might guess, these examples have induced attempts to extend the theory of compact semigroups and scattered, partial success has resulted. One notable instance of success is the extension of (2.8) to the locally compact case in [10], that is, a locally compact semigroup which is a group is a topological group. (A space X is locally compact iff each point x of X is contained in an open set V_x whose closure, V_x^*, is compact). R is locally compact under the usual topology and so is a topological group under addition. However, this extension of (2.8) to the locally compact case implies that R with the topology given in (g) is not locally compact, an interesting way to obtain this purely topological result.

So far the noncompact semigroups we have listed have all been abelian. Let us turn our attention to a nonabelian example. In Section 2 of Chapter 3 a nonabelian, compact semigroup D was discussed. If we remove one element from D, namely, (0,0), it is not hard to verify that the reduced set is not compact and yet it is a semigroup--one must be careful to maintain algebraic closure.

Actually, noncompact semigroups are profusely abundant if the discrete topology is used. For, any infinite semigroup with the discrete topology, clearly, fails to be compact.

Section 5. Threads

This section discusses semigroups which topologically look like a closed interval [a,b] of the real line R. To characterize an interval topologically we first need the notion of a "cut point": p is a <u>cut point</u> of a connected space X iff X\p is not connected. Then, any space--such as a closed interval--which is a compact, connected, Hausdorff space having the property that every point is a cut point except two points--called endpoints--is termed an <u>arc</u>. In this terminology, this section deals with those semigroups, often called <u>threads</u>, whose under-lying space is an arc. Our discussion will center upon the concept of an <u>I-semigroup</u>, a thread in which one end-point is a zero and the other is an identity. This semi-group is well named because in [11] William M. Faucett proved that such a thread having no interior idempotents must be homeomorphic to the interval [0,1], often designated by I.

Several examples play a paramount role in the theory of I-semigroups and to them we now turn our attention: (1) a <u>unit thread</u> is a semigroup which is isomorphic to [0,1] using ordinary multiplication. Of course, a unit thread is an I-semigroup.

(2) Another important I-semigroup is a <u>nil thread</u>, a semigroup isomorphic to [1/2, 1] with multiplication * defined by x*y = maximum {1/2, the ordinary product xy}.

It is of interest to note that a nil thread is the same as the Rees quotient semigroup I/[0, 1/2] where I represents a unit thread. (We shall postpone direct verification of the continuity of * until after we have discussed the "min" thread in (3).) One way to distinguish a nil thread N from a unit thread is to notice that every subsemigroup generated by a single element in N, except 1, contains the zero, namely, 1/2 in this case. We say, algebraically, that an element x in a semigroup S is <u>nilpotent</u> iff for some integer n $x^n = 0$, the zero. (If every element of S is nilpotent, S is said to be <u>nil</u>. Since all elements of N, except 1, are nilpotent this point is emphasized in the name, nil thread.) Therefore, N contrasts sharply with the unit thread I where the only nilpotent element is the zero.

I and N are more than merely examples of I-semigroups. For, together they characterize all I-semigroups containing just two idempotents: Let T be an I-semigroup containing just two idempotents. It can be shown that if T has no (nonzero) nilpotent elements, then T is isomorphic to I; if T does have at least one nonzero nilpotent element, then it is isomorphic to N. However, I and N do not typify all I-semigroups as the next example shows:

(3) Lastly, we consider a third type of I-semigroup. A <u>min thread</u> is a semigroup whose underlying arc is a closed interval [a,b] of real numbers and its multiplication # is defined by x#y = minimum {x,y}. The distinguishing feature of a min thread is that every element is an

120

idempotent.

To show that # is continuous we shall consider a more
general situation, namely, a "min" function with domain
R × R and then regard # as a restriction. Letting q =
min {x,y}, consider any open set V containing q. (We
proceed to make use of (1.31).) Since it is easy to
see that the range of this general min function is R,
one may find a positive number ε such that the interval
(q - ε, q + ε) is contained in V. Select the open set
U = (x - ε, x + ε) × (y - ε, y + ε). For any point (a,b)
in U it is not difficult to verify that min {a,b} is in V.
Therefore, min(U) ⊆ V and so, by (1.31), this min function
is continuous. #, being a restriction of min, is also
continuous by (1.25).

At this point we return to verify the continuity of
the nil thread multiplication *. The function c:R × R → R
defined by c(x,y) = 1/2 is continuous, as shown by a simple
argument similar to that used for the constant function of
(1.32). Then by (2.5) we may construct a continuous
function g = (c,m) where m represents ordinary multiplication. In addition, the function h, defined by h(x,y) =
max {x,y} with domain R × R is continuous, the proof being
analogous to that for the min function. Consequently, the
composition function hg:R × R → R defined by hg(x,y) =
max{1/2, xy} is continuous by (1.2). Finally, the
multiplication *, being a restriction of hg, is continuous

by (1.25).

It is remarked that all I-semigroups are abelian.
However, not all threads have this property. Consider
in the plane the set L = ({0} × [0,1]) $\cup$ ([0,1] × {0}) and
define a multiplication $ on L by (x,y)$(x',y') =
(xx',xy' + y).

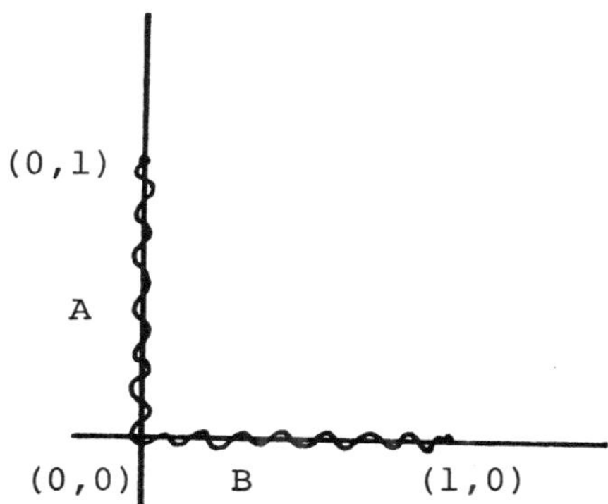

Figure 7.7   L = A $\cup$ B

To show the continuity of $, one may use (2.5) with
$ = (f_1,f_2). The function f_1 is continuous, by (1.2),
since it is the first projection of the semigroup product
multiplication, that is, $f_1 = \pi_1 P$.

The continuity of f_2 requires more explanation.
Actually, $ is defined in a coordinate manner, i.e.,
p$q has meaning when p and q, being (restricted) points
in the plane, are expressed as p = (x,y) and q = (x',y').
We make this observation in order to note that the composite
function $\pi_1 \pi_1 (p,q)$ = x is continuous by (1.2) as also is
$\pi_2 \pi_2 (p,q)$ = y'. Then by (2.5) the function g = ($\pi_1 \pi_1$, $\pi_2 \pi_2$)

is continuous and so the composition function mg, where m is ordinary multiplication, is continuous. Also, the composition function $\pi_2\pi_1(p,q) = y$ is continuous. Therefore, by (2.5), the function $h = (mg, \pi_2\pi_1)$ is continuous. Finally, the composition function $ah = xy' + y$, where a represents ordinary addition, is continuous and, of course, f_2 is ah.

The associativity of \$ is easy to verify and the details are left to the reader. It is clear that \$ is not an abelian function.

It is noted that (1,0) is an identity for L. In addition, every element on the vertical portion, the set A in Figure 7.7, is idempotent. However, L has no zero.

APPENDIX

CONTINUITY OF ORDINARY ADDITION AND MULTIPLICATION

These two proofs have been entered as an appendix for two reasons: (1) The reader probably has encountered them in calculus courses. (2) Implicitly, the nature of the proofs is metric, a concept this book does not deal with directly. Another way to phrase the second reason is to say that some commonly assumed real number properties, which are metric in nature, are utilized in these proofs.

A.1. Proposition. Ordinary arithmetic addition of real numbers is a continuous function.

Proof. Let $\bar{a}:R \times R \to R$ be defined by $\bar{a}(x,y) = x + y$. We proceed to verify that $\bar{a}$ is continuous, making use of an equivalent definition of continuity given in the (1.31) Exercise.

Upon reflection we note that $\bar{a}$ is surjective. Therefore, selecting any point (p,q) in the domain of $\bar{a}$, one finds that any open set W containing $p + q$ contains a base member of the topology of R, namely, an open interval V, which also contains $p + q$. If we find an open set U in $R \times R$ with the property $\bar{a}(U) \subset V$, then clearly $\bar{a}(U) \subset W$. For this reason, in utilizing (1.31) it suffices to consider only base members containing $p + q$.

Let V be any open base member containing $p + q$ and suppose the width of V is ε. Therefore, for any other

point (x,y) with value x + y in V we know that $|\bar{a}(x,y) -$
$\bar{a}(p,q)| < \varepsilon$. Moreover, this inequality holds for all
those points (x,y) satisfying $|x - p| + |y - q| < \varepsilon$
because $|\bar{a}(x,y) - \bar{a}(p,q)| = |(x + y) - (p + q)| =$
$|(x - p) + (y - q)| \leq |x - p| + |y - q|$.

Now select δ to be any positive number less than $\varepsilon/2$
and consider the cartesian product U of the two open
intervals $(p - \delta, p + \delta)$ and $(q - \delta, q + \delta)$, i.e.,
$U = (p - \delta, p + \delta) \times (q - \delta, q + \delta)$, noting that U is
open by the definition of a product topology (because it
is a base member). For any point (x,y) in U one finds
$|\bar{a}(x,y) - \bar{a}(p,q)| < \varepsilon$ because $|x - p| < \delta$ and $|y - q| < \delta$.
In other words, $\bar{a}(U) \subset V$ and, consequently, by (1.31) $\bar{a}$
is continuous. $\square$

A.2. <u>Proposition</u>. Ordinary arithmetic multiplication
of real numbers is a continuous function.

Proof. Let $\bar{m}$ denote the multiplication function. As
in (A.1), we will show continuity by means of (1.31).

Let (s,t) be any point in the domain of $\bar{m}$. Once again
it suffices to consider only base members containing st.
So, let V be an open base member containing st and suppose
the width of V is α. Therefore, for any other point (x,y)
with value xy in V we know that $|\bar{m}(x,y) - \bar{m}(s,t)| < \alpha$.
The set U required in (1.31) will, again, be of the form
$(s - \delta, s + \delta) \times (t - \delta, t + \delta)$, but the process of selecting
an appropriate δ is far more subtle than in (A.1). Indeed,

this selection of δ is the only real distinction between the proofs of the (A.1) and (A.2) Propositions.

For a while we leave δ undetermined except to say that we will be considering points (x,y) satisfying $|x - s| < \delta$ and $|y - t| < \delta$, i.e., x in $(s - \delta, s + \delta)$ and y in $(t - \delta, t + \delta)$. In the following sequence of equalities and inequalities we use the common device of adding and subtracting a quantity, in this case sy, and the well-known facts concerning absolute values, $|ab| = |a||b|$ and $|a+b| \leq |a| + |b|$. We find $|\overline{m}(x,y) - \overline{m}(s,t)| = |xy - st| = |xy - sy + sy - st| \leq |xy - sy| + |sy - st| = |y||x-s| + |s||y-t| < |y|\delta + |s|\delta < |t+\delta|\delta + |s|\delta \leq (|t| + \delta)\delta + |s|\delta = \delta^2 + \delta(|t| + |s|)$. To summarize, $|\overline{m}(x,y) - \overline{m}(s,t)| < \delta^2 + \delta(|t| + |s|)$ for those points (x,y) satisfying $|x - s| < \delta$ and $|y - t| < \delta$.

At last we are in a position to indicate how to explicity determine the necessary value for δ: From what we have just seen, it is clear that in order for the inequality $|\overline{m}(x,y) - \overline{m}(s,t)| < \alpha$ to hold, it suffices to have $\delta^2 + \delta(|t| + |s|) < \alpha$. Consequently, δ may be chosen by using the quadratic formula to solve the equation $\delta^2 + (|t| + |s|)\delta - \theta = 0$ where θ is any positive number less than α, say for example $\alpha/2$.

As already mentioned, this selected value of δ yields an open set U and the preceding remarks verify, in a point-wise fashion, that $\overline{m}(U) \subset V$. Therefore, by (1.31), $\overline{m}$ is continuous. $\square$

SEMIGROUP BIBLIOGRAPHY
Chapters 2-7

(Topological) Semigroups Texts

1. Hofmann, K. H. and Mostert, P. S., _Elements of Compact Semigroups_, Merrill Books, Inc., Columbus, Ohio 1966.

2. Paalman- De Miranda, A. A., _Topological Semigroups_, Mathematical Centre Tracts, Amsterdam, 1964.

Algebraic Semigroup Texts

3. Clifford, A. H., and Preston, G. B., _The Algebraic Theory of Semigroups_, Math Surveys No. 7, Amer. Math. Soc., Providence, 1961.

4. Ljapin, E. S., _Semigroups_, Trans. Math. Monographs, Vol. 3, Amer. Math. Soc., Providence, 1963.

Expository Articles

5. Day, J. M., "Expository lectures on topological semigroups", Chapter 10 in _Algebraic Theory of Machines, Languages, and Semigroups_, edited by M. A. Arbib, Academic Press, New York City, 1968, pp. 269-296.

6. Mostert, P. S., "The structure of topological semigroups-revisited", Bull. Amer. Math. Soc., Vol. 72 (1966), pp. 601-618.

7. Wallace, A. D., "The structure of topological semigroups", Bull. Amer. Math. Soc., Vol. 61 (1955), pp. 95-112.

Research Articles

8. Arens, R., "Linear topological division algebras", Bull. Amer. Math. Soc., Vol. 53 (1947), pp. 623-630.

9. Bastida, J. R., "Grupos y homomorfismos asociados con un semigrupo, I", Bol. Soc. Mat. Mex., Vol. 8 (1963), pp. 26-45.

10. Ellis, R., "A note on the continuity of the inverse",
        Proc. Amer. Math. Soc., Vol. 8 (1957), pp. 372-373.

11. Faucett, W. M., "Compact semigroups irreducibly connected
        between two idempotents", Proc. Amer. Math. Soc.,
        Vol. 6 (1955), pp. 741-747.

12. Wallace, A. D., "The Rees-Suschkewitsch structure theorem
        for compact simple semigroups", Proc. Nat. Acad. Sci.
        U.S.A., Vol. 42 (1956), pp. 430-432.

13. _______________, "Relative ideals in semigroups. I",
        Colloq. Math., Vol. 9 (1962), pp. 55-61.

14. _______________, "Relative ideals in semigroups. II",
        Acta Math., Vol. 14 (1963), pp. 137-148.

Topological Group Text

15. Pontryagin, L. S., _Topological Groups_, Gordon and Breach,
        New York City, 1966.

Unpublished Lecture Notes on Semigroups

Wallace, A. D., "Project mob", University of Florida, Gaines-
        ville, 1965.

Articles on Examples of Finite (algebraic) Semigroups

Forsythe, G. E., "SWAC computes 126 distinct semigroups of
        order 4", Proc. Amer. Math. Soc., Vol. 6 (1955),
        pp. 443-445. (It is noted that this article also
        lists the 4 semigroups of order 2 and the 18 semi-
        groups of order 3.)

Plemmons, R. J., "There are 15973 semigroups of order 6",
        Math. Algorithms, Vol. 2 (1967), pp. 2-17. (Also
        see Vol. 3 (1968), page 23, for a Remark by the
        Editor on this article.)

Tetsuya, K. T., T. Hashimoto, T. Akazawa, R. Shibota, T. Inue,
        T. Tamura, "All semigroups of order at most 5", Journal
        of Gakugei, Takushima University, Vol. 6 (1955),
        pp. 19-39.

ANSWERS TO SELECTED EXERCISES

Chapter I

1.22. An example of an "element argument" proof is the proof of the 1.3 Proposition.

1.24. It is the indiscrete topology.

1.26. First, f being continuous implies that $f(A^*) \subset f(A)^*$. Secondly, A^* is compact because X is compact and, consequently, $f(A^*)$ is compact since f is continuous. Then, since Y is Hausdorff, $f(A^*)$ is closed by the 1.8 Proposition. Therefore, $f(A)^* \subset f(A^*)^* = f(A^*)$. That is, it is also true that $f(A)^* \subset f(A^*)$.

1.27. See page 95 of <u>General Topology</u> by J. L. Kelley for the proof of this exercise.

1.28. The onto portion of the premise is needed so that f^{-1} will have Y as its domain. It then suffices to show that f^{-1} is continuous: Let A be closed in X. It is easy to show that $(f^{-1})^{-1}(A) = f(A)$ because f is 1-1. Since X is compact, f is continuous, and Y is Hausdorff, it follows (as in the proof of 1.26) that $f(A)$ is closed.

1.29. (a) Let S be the circle $x^2 + y^2 = 1$ in the plane R^2. Let T be the perimeter of the rectangle bounded by the lines $x = 1$, $y = 1$, $x = -1$ and $y = -1$. $S \cap T = \{(1,0),(0,1),(-1,0),(0,-1)\}$. Clearly, these four points are not connected.

Another example is given in the figure which follows:

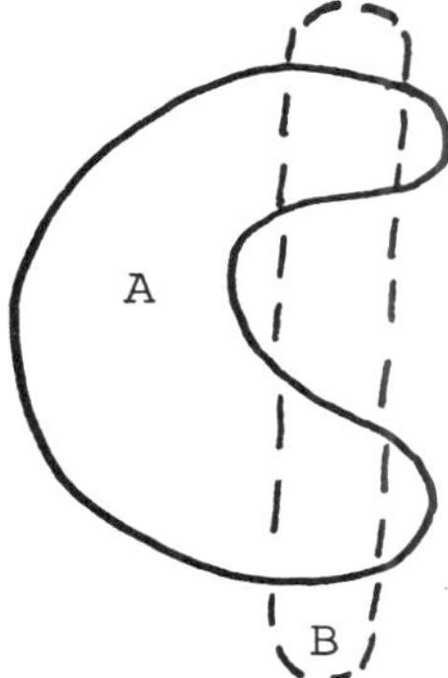

A and B are connected whereas A ∩ B is not connected.

1.32.   (a) Let A be any set in R. c(A*) = k. Also, c(A) = k which is a closed singleton set. That is, c(A*) = [c(A)]* and so c is continuous by the 1.1(b) Proposition.

1.32.   (b) This result follows from the basic definition of a continuous function because the inverse image of each open set V is V itself.

1.33.   Consider the constant function c:R → R defined by c(x) = k. Let D be the union of the intervals (0,1) and (10,11). c(D) = k is connected.

Chapter II.

2.13.   a and b. $\mathscr{S}$ is a semigroup under addition and it is also a semigroup under multiplication. However, it is not a group under either operation because in both cases, using the remark in the 2.12 exercise, "inverses" (i.e., elements g' such that gg' = g'g = e) do not exist. (The reader should determine what the identity, which does exist, is in both cases.)

2.15.　$\mathcal{N}$ is a semigroup under matrix addition. However, it is not a group because it lacks an identity.

2.16.　Consider the product of the following two matrices in $\mathcal{L}$.

$$\begin{pmatrix} 1 & -1 \\ -1 & 1 \end{pmatrix} \begin{pmatrix} 1 & 1 \\ 1 & 1 \end{pmatrix}$$

2.17.　$\mathcal{I}$ is a semigroup under multiplication. However, it is easy to verify that no solution exists to the equation AX = B where A = $\begin{pmatrix} 1 & 1 \\ 1 & 1 \end{pmatrix}$ and

B is the identity matrix $\begin{pmatrix} 1 & 0 \\ 0 & 1 \end{pmatrix}$.

Thus, by the 2.12 exercise, $\mathcal{I}$ is not a group.

2.18.　Remarks in the answer to 2.17 apply here also.

2.19.　Answer is similar to that for 2.15.

2.20.　The following sets are groups under addition:

R, R*, Z, E, {0}

The following sets are groups under multiplication:

R, R*, R(0), R*(0), {0}, {1}.

The reader should provide examples to show that the other sets listed at the beginning of this chapter are not groups.

2.21.　R<0> is a semigroup under addition but it is not a semigroup under multiplication because the product of any two negative numbers is positive.

2.23.   Consider the multiplicative semigroup R(0), that is,
        [0,∞) with the usual (relative) topology. (0,∞) is
        a maximal subgroup which, clearly, is not closed
        (because {0} is closed and so R(0)\{0} = (0,∞) is
        open).

Chapter III.

3.17.   The proof is similar to the first part of the proof
        for (3.2.c).

3.18.   (a) Any group is a simple semigroup.

        (b) [0,1). This set is open because it is the
            intersection of [0,1] and $(-\infty,1)$.

        (c) Let the space be [0,1]. Then $[0,\gamma]$ for any real
            number γ such that $0<\gamma<1$ is a proper closed
            ideal which is not a kernel.

        (d) Let S = (0,1). Each interval $(0,\gamma)$, where γ
            is a real number and $0<\gamma<1$, is an ideal. However,
            there is no minimal ideal.

        (e) Let S = the set of real numbers, R, under multi-
            plication with the usual topology. The kernel
            is zero.

        (f) Let S = {-1,0,1} under multiplication with the
            discrete topology. {0} is then an open minimal
            ideal. (In fact, any semigroup having zero and
            using the discrete topology will have an open
            minimal ideal.)

3.19.   We will show that if L is any left ideal, then
L = G: Of course, L $\subseteq$ G. Also, G = GL since G
is a group, and GL $\subseteq$ L since L is a left ideal.
This implies that G $\subseteq$ L. Therefore, L = G and a
group cannot have any proper left ideals.

3.20.   Let I be an ideal which is a group. If J is any
other ideal, then I $\cap$ J is an ideal by the 3.2.c
Proposition. I $\cap$ J is in the group I so that, by the
3.19 Exercise, I $\cap$ J = I.   Therefore I $\subseteq$ J and so
I is a minimal ideal.
Let G be the maximal subgroup which contains I.
I is also an ideal of G. Let e be the identity
element in I and G. Then G = Ge $\subseteq$ GI $\subseteq$ I. Therefore,
I = G.

3.21.   That AB is compact follows from the 1.10 and 1.11
Theorems. That AB is connected follows from both the
last paragraph of Chapter I (before the Exercises)
and the 1.19 Theorem.

3.22.c. $\{(1/2, 0), (-1/2, 0)\}$ is not a right ideal.

3.23.   It is easy to see that SaS is an ideal. With this
observation, the proof is similar to part (a) of 3.9.

3.24.   The 3.9 Proposition is needed for all three results.
In addition, the 3.4, 3.10 and 1.8 Propositions are
needed to prove one or more of these three results.

3.25.   We will show that S is a group and then the result
will follow from the 2.8 Theorem: First, S = K

because e is in K and S = Se ⊂ SK ⊂ K. Since K is the union of minimal left ideals, e is in a minimal left ideal; call it N. It follows that S = Se ⊂ SN ⊂ N and, therefore, K = S = N and so K is a minimal left ideal. By (3.9) N = Sa for any a in N and thus, S = Sa for all a in S. In a similar fashion, S = K is a minimal right ideal and S = aS for all a in S. By definition, S is a group.

3.26. Let M be any minimal left ideal of S. We will show that M = Na for any a in M: By (3.9) M = Sa for any a in M. Therefore, Na ⊂ Sa = M for any a in M. However, noting that Na is a left ideal, then M = Na, for any a in M, because M is a minimal left ideal and cannot properly contain any left ideals.

3.27. (a) K = N_1 ∪ N_2 is not connected and, since S is connected, K should be connected by the 3.12 Theorem.

(b) N_2 ∩ L_1 is a left ideal and it is contained in both L_2 and N_2. This containment contradicts the minimality of N_2 in L_2.

(c) The distinct minimal ideals, N_1 and N_2, should be disjoint by (3.8).

3.28. Let J be an ideal of S. K ∩ J is an ideal, by (3.2), and K ∩ J ⊂ J. Also, K ∩ J ⊂ K which implies K ∩ J = K because K is a minimal ideal. Consequently, K ⊂ J.

3.29. The set (0,a) for any a in (0,1) is an ideal of

134

S = (0,1). Note that the intersection of these
ideals is empty. But, by 3.28, K is in each of these
ideals and therefore K is contained in their intersection. Consequently, K is empty and doesn't exist.

3.30. Suppose S is simple. Since SaS is an ideal of S,
it follows immediately that SaS = S. Conversely,
suppose S is not simple and let I be a proper
ideal of S. For any a in I, $SaS \subset SIS \subset I \neq S$.
That is, $SaS \neq S$ when a is in I.

3.31. This result is similar in statement and proof to
3.28. Let J be an ideal and N be a minimal left
ideal. In this case, note that JN is a left ideal
and $JN \subset N$. Since N is minimal, JN = N. Also,
$JN \subset JS \subset J$. Therefore, $N \subset J$.

Chapter IV.

4.18. (a) By (3.16) K is the union of its left minimal
ideals. Hence, K = S and, using (3.28), S is
simple.

(b) Since S is compact, a minimal left ideal and
a kernel K exists by (4.1) and (4.2). Further,
since S is simple, S = K. Then, by (3.16),
S = K is the union of its left minimal ideals.

4.19. By (4.1) S contains a minimal left ideal. Then,
by (3.16) S has a kernel.

4.20. (Note that this exercise is asking us to prove the
4.2 Theorem without using (4.1).) Consider the

collection $\mathscr{C}$ of all closed ideals. $\mathscr{C}$ is not empty because S is an ideal and it is closed by (1.8). Since, for any ideals I and J, $IJ \subset (I \cap J)$, it is easy to extend this fact to show that the product of a finite number of ideals, say $I_1 I_2 \ldots I_N$, is contained in each of these N ideals and so $\mathscr{C}$ has the finite intersection property. Therefore, since S is compact, all the sets in $\mathscr{C}$ have a nonempty intersection; call this intersection F. F is an ideal by (3.2.b). It remains to show that F is minimal: Suppose J is an ideal contained in F and let a be any element in J. Then $SaS \subset SJS \subset J \subset F$. But SaS is an ideal and it follows easily that it is closed because S is compact. Therefore, SaS is a member of $\mathscr{C}$ and $F \subset SaS$. That is, SaS = F and so J = F which shows that F is a minimal ideal. One may apply one of DeMorgan's laws to the collection $\mathscr{C}$ to verify that F is closed.

4.21. By (1.5) S × S is Hausdorff. $\theta(S \times S)$ is a subset of S × S by the definition of θ and consequently, by (1.23), $\theta(S \times S)$ is also Hausdorff. Then, using (1.6) with $Y = \theta(S \times S)$, $\Delta(\theta)$ is closed.

4.22. 0(x) is abelian because of the associative law. This result now follows from (4.7).

4.24. The kernel K is a group by (4.9) because the semigroup S is abelian. K is compact by (3.12) because

136

S is compact. Then, by (2.8), K is a topological
group.

4.25. The semigroup S being compact implies that it contains
a kernel K. S being simple implies that S = K.
S being abelian implies K = S is a group by (4.9).

4.26. $\Gamma(f) = f$ because in a Hausdorff space single points
are closed.

4.27. Let x be an element in the semigroup S. $\Gamma(x)$ is
closed and, by (1.7), $\Gamma(x)$ is compact because S
is compact. Therefore, using (4.10), S has an
idempotent.

4.28. Here are two examples: (a) Let S = [0,1] under
multiplication and with the usual topology for real
numbers. 1 is an identity. (Note that S is compact.)
[0, 1/2] is a compact subsemigroup which is not a
group. (b) Let S = R(0), the set of nonnegative
real numbers. Again, use multiplication and the
usual topology. (Note that this semigroup is not
compact.) Again [0, 1/2] is a compact subsemigroup
which is not a group.

4.30. (a) Let the operation in the group be denoted by
multiplication. First, we will verify that the
identity e is in the semigroup S: Let a be any
element in S. Since S is finite, $a^n = a$ for some
integer n. So, $a \cdot a^{n-1} = a$ and $a^{n-1} = e$ because
in a group the identity is unique. Therefore, e

is in S. Now, a = ae is in aS and, consequently,

S ⊆ aS. Therefore, aS = S. Similarly, S = Sa and

S is a group by definition.

(b) The result in part (a) was independent of the

underlying topological properties of the group G

and its subsemigroup S. However, the same result

could be obtained by using (4.14) once we observe

that every finite semigroup is both compact and

closed. In particular, in this case, G is compact

and S is closed.

4.31. $T(A) \subseteq A$ implies $[T(A)]^* \subseteq A^* = A$. Since $[T(A)]^*$ is

an ideal, by (3.1), which is contained in A, then

$[T(A)]^* \subseteq T(A)$ by the definition of $T(A)$. Therefore,

$T(A)$ is closed.

4.32. Let a be any element in (1/4, 1/2]. Let $I = [0,1/2] \backslash a$,

that is, I is [0, 1/2] with the one element a

deleted. I is a maximal proper ideal. (Consequently,

there are an infinite number of maximal proper ideals.)

4.33. Let S = R, the set of real numbers, with multiplication

and the usual topology. The singleton set, {0},

is a maximal proper closed ideal of S.

4.34. Let S be (0,1) under multiplication and with the

usual topology. S has no proper maximal ideals.

(Note: S is not compact.)

4.35. In [0,1] with multiplication and the usual topology,

the subset $[0,1] \backslash \{1\}$, i.e., [0,1] with the one point

1 deleted, is a maximal proper ideal. (Note that this ideal is dense in [0,1].)

4.36. (a) The fact that closed sets are disjoint does not imply that their union is disconnected. For example, each point in an open interval (a,b) is closed and, consequently, any two points are separated. That is, not only are two points disjoint but they are not connected also. However, the union of these points is a connected interval. (Moreover, since (a,b) is open, this example also illustrates that the union of closed sets, namely, the singleton sets, need not be closed.)

4.36. (b) Consider the subset D of R^2 defined in section 3.2 and illustrated in Figure 3.1. The product of two elements, $(x,y)*(a,b)$ is defined by the second column in the product of the matrix multiplication

$$\begin{pmatrix} 1 & x \\ 0 & y \end{pmatrix} \begin{pmatrix} 1 & a \\ 0 & b \end{pmatrix} \quad .$$

Clearly, D is connected and, also, it is compact (since $|x| + |y| \leq 1$). The points $\{(-1/2, 0)\}$, $\{(0, 0)\}$ and $\{(1/2, 0)\}$ are three distinct minimal left ideals. (It is easy to verify that (0, 0) is not a right ideal by finding the product of

$$\begin{pmatrix} 1 & 0 \\ 0 & 0 \end{pmatrix} \begin{pmatrix} 1 & x \\ 0 & y \end{pmatrix} \quad .$$ In fact, every point on the line segment [-1/2, 1/2] is a minimal left ideal

and this interval, $[-1/2, 1/2]$, is the kernel, by
(3.16). (The reader should verify that $[-1/2, 1/2]$
is a right ideal.)

Chapter V.

5.9.   Let f be an idempotent in S, that is, $f^2 = f$. Then,
$h(f)h(f) = h(f^2) = h(f)$ and so $h(f)$ is an idempotent
in S'.

5.10.   The reader can easily verify that a semigroup S is
simple if and only if S is its own minimal ideal.
By part (b) of (5.3), $p(S) = T$ is a minimal ideal.
Therefore, T is also simple by the initial remark.

5.11.   Suppose, on the contrary, that x is not in W_1.
Then x is in $n(S\setminus U)$, that is, there is an element
a in $S\setminus U$ such that $n(a) = x$. Thus, a is in $n^{-1}(x)$
$\subset U$. So, $a \in S\setminus U \cap U$, which is contradictory.
Hence, our original assumption that $x \notin W_1$ is incorrect.

5.12.   (a) 1. The product of Hausdorff spaces is Hausdorff, i.e.,
the (1.5) Theorem.   2. $S/\mathcal{C}^{\circ}$ is Hausdorff.

5.12.   (b) One should keep the diagram, given in the proof
of (5.8), in mind when reading this argument: Let
(A,B) be in $\mu^{-1}(Z)$. Since $n \times n$ is a surjective
function, $(A,B) = n \times n (a,b)$ for some $a \in A$, $b \in B$.
Further, from the diagram, $n(m(a,b)) = \mu(n \times n(a,b))$.
Therefore, $n(m(a,b)) = \mu(A,B)$ is in Z and, consequently,
(a,b) is in $m^{-1}n^{-1}(Z)$. As a result, $(A,B) = n \times n(a,b)$

is in n × n$[m^{-1}n^{-1}(Z)]$. To prove the other set
inclusion, let (A,B) be in n × n$[m^{-1}n^{-1}(Z)]$. Then
there exists an (a,b) in $m^{-1}n^{-1}(Z)$ such that
n × n(a,b) = (A,B). Since n(m(a,b)) is in Z, it
follows that μ(n × n(a,b)) is in Z by the definition
of μ. That is, n × n(a,b) = (A,B) ε $μ^{-1}(Z)$.

5.13. I "induces" the congruence I × I ∪ Δ. I is compact
because it is closed and S is compact. Therefore,
I × I is compact and, since S × S is Hausdorff,
I × I is then closed by (1.8). Δ is closed by (1.6)
because S is Hausdorff. Therefore, the union of two
closed sets is closed.

5.14. Six different examples will follow. The first two
are non-compact and the last four are compact.

(1) Let S_1 be the open interval (0,1) under multiplication. To form the quotient semigroup use
the ideal I = (0,a) for any a in (0,1).

(2) Consider Z(1), the positive integers, under
multiplication. Use the ideal E(2), the positive
even integers, to form the quotient semigroup.
(The reader should determine what are the elements
of Z(1)/E(2).)

S_3 - S_6 are semigroups each having four elements
and defined by the "multiplication" tables given
below. All four semigroups are compact (using the
discrete topology). In S_3, letting I_3 = {0,1,3},

then S_3/I_3 is of order 2. If we let $I_4 = \{0,1\}$ in S_4, then S_4/I_4 is of order 3. In both S_5 and S_6 define $I = \{0,2\}$. In S_5/I, the set $\{\overline{1},\overline{3}\}$ forms a subgroup; however, in S_6/I the set $\{\overline{1},\overline{3}\}$ doesn't even form a semigroup. Consequently, the algebraic structures of the quotient semigroups S_5/I and S_6/I are substantially different. (The ideal is the zero.)

S_3

*	0	1	2	3
0	0	1	1	3
1	1	3	3	0
2	1	3	3	0
3	3	0	0	1

S_4

*	0	1	2	3
0	0	1	1	1
1	1	0	0	0
2	1	0	0	0
3	1	0	0	0

S_5

*	0	1	2	3
0	0	0	2	2
1	0	1	2	3
2	2	2	0	0
3	2	3	0	1

S_6

*	0	1	2	3
0	0	0	2	2
1	0	1	2	3
2	2	2	0	0
3	2	3	0	0

Chapter VI.

6.11. This result is almost immediate: Let a and b be elements of $A^{[-1]}A$. Then $A(ab) = (Aa)b \subset Ab \subset A$ and so ab is also an element in $A^{[-1]}A$. Consequently, $A^{[-1]}A$ is a subsemigroup.

142

6.12. First let us verify the two facts asserted in the hint: (1) To show that $X^{[-1]}Y = \cap \{a^{(-1)}Y; a \in X\}$ first suppose that $b \in X^{[-1]}Y$, i.e., $Xb \subset Y$. Consequently, for each $a \in X$, $ab \subset Y$; in other words, b is in $a^{(-1)}Y$ for each $a \in X$. Therefore, b is in $\cap \{a^{(-1)}Y; a \in X\}$. Conversely, let c be in $\cap \{a^{(-1)}Y; a \in X\}$. Then $ac \subset Y$ for each a in X and so c is in $X^{[-1]}$.

(2) The following interesting proof shows that the function $1a$ is continuous: Let f_1 be the constant function, $f_1(s) = a$, and f_2 be the identity function, $f_2(s) = s$. One can easily verify that both f_1 and f_2 are continuous by viewing the proof of exercise (1.32). (The proof of (1.32a), that a constant function is continuous, applies not only to R but also to any topological space in which singleton sets are closed and this is true in all Hausdorff space; the proof of (1.32b), that the identity function is continuous, is even more general than (1.32a) and holds for any topological space.) Then, by the (2.5) topological result, the function $f(s) = (f_1(s), f_2(s)) = (a,s)$ is continuous. Finally, the composition function $m(f(s))$ is continuous by (1.2) because the semigroup multiplication m is also a continuous function. This composition is the same as the function $1a$, as may be seen from the following diagram:

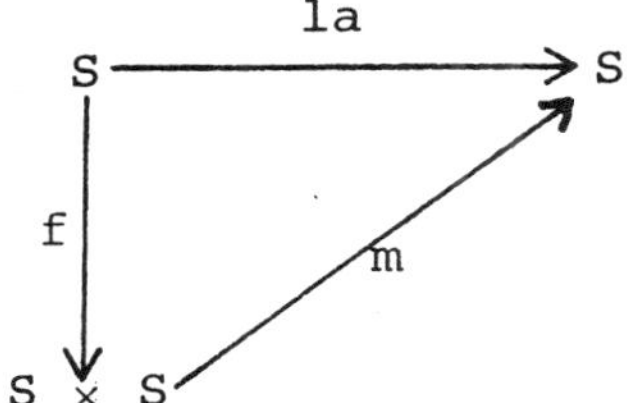

One more result will be needed to prove (6.12):
The intersection of any number of closed sets is a
closed set; this fact follows from the basic definition of a topology and DeMorgan's Laws.
Now, since for each a in X la is a continuous
function and Y is closed, then $(la)^{-1}(Y) = a^{(-1)}Y$
is closed. Therefore, in view of the previous
remarks, $\cap\{a^{(-1)}Y;\ a\ \varepsilon\ X\} = X^{[-1]}Y$ is also closed.

6.13. $L(y) = y \cup Sy = y \cup m(S,y) = y \cup m(S \times \{y\})$. Since S
and $\{y\}$ are compact and the semigroup operation m
is continuous, then $m(S \times \{y\})$ is compact. It
follows that $m(S,y)$ is closed because S is Hausdorff
by (1.8). Finally, the union of two closed sets is
closed and thus $L(y)$ is closed.

6.14. Suppose that (γ,t) is in $\mathscr{L}$. This means $\gamma \cup S\gamma =$
$t \cup St$. It follows that $\gamma x \cup S\gamma x = (\gamma \cup S\gamma)x = (t \cup St)x =$
$tx \cup Stx$. Consequently, $(\gamma x, tx)$ is in $\mathscr{L}$ and $\mathscr{L}$ is
a right congruence.

6.15. First we will show that if $x\ \varepsilon\ L_a(T)$, then $x\ \varepsilon\ L_a(S)$:
Suppose that $(x,a)\ \varepsilon\ \mathscr{L}(T)$. This means $x = a$ or

144

x ε Ta. If x ≠ a, then x ε Sa, say x = s'a for some

s' ε S, because T ⊂ S. Hence, if x ≠ a, x ∪ Sx =

(s'a ∪ Ss'a) ⊂ Sa. Therefore, it follows easily

that L(x,S) ⊂ L(a,S). Similarly, L(a,S) ⊂ L(x,S)

and so x ε L_a(S).

Now suppose y ε L_a(S). Clearly, y is also in some

L_a(T)-class. But then, by the preceding argument,

all of that L_a(T)-class is in L_a(S). Therefore,

considering each y in L_a(S) gives us that L_a(S) is

the union of L_a(T)-classes.

6.16. Let g ε G. Since G is a group, gG = Gg = G and it

follows quickly that H(g,G) = H(g', G) for all g,g'

in G. Therefore, G ⊂ H_g(G). Conversely, suppose

g ε G and a ε H_g(G). Then (a ∪ aG) = (g ∪ gG) = G

so that a ε G. That is, H_g(G) ⊂ G.

6.17. (a) By (6.9) H_e is a maximal subgroup. (2.11) states

that a maximal subgroup is closed when S is compact.

Therefore, H_e is closed.

(b) Let $\mathscr{G}$ be the union of all subgroups of S. Let

us verify that $\mathscr{G}$ = ∪ {H_e; e ε E}: Suppose a is a

member of a subgroup A in $\mathscr{G}$. A is contained in a maximal

subgroup which has the same identity, say e, as A and so

a ε H_e by (6.9). Conversely, let x ε ∪{H_e; e ε E}. Then x

is in a maximal subgroup, again by (6.9). Hence, x ε $\mathscr{G}$.

This exercise now follows from (6.10) which says that $\mathscr{G}$

is closed when S is compact.

6.18.　(a) Recall that π_2 is a projection map of the product space $S \times S$ upon S and that $L_a = \{x; (x,a) \in \mathscr{L}\}$. It is easy to verify that $L_a = \pi_1(\pi_2^{-1}(a) \cap \mathscr{L})$: Suppose $x \in L_a$. Then (x,a) is in both $\pi_2^{-1}(a)$ and $\mathscr{L}$ and so $x \in \pi_1(\pi_2^{-1}(a) \cap \mathscr{L})$. Now suppose $t \in \pi_1(\pi_2^{-1}(a) \cap \mathscr{L})$. Then (t,a) is in $\mathscr{L}$ which implies that $t \in L_a$. (The associated "picture" in $S \times S$ is also convincing and the reader is encouraged to draw it.) $\mathscr{L}$ is closed when S is compact, by (6.6), and $\pi_2^{-1}(a)$ is also closed because π_2 is continuous and singleton sets are closed in a Hausdorff space. Therefore, the conclusion follows that L_a is closed in view of (2.7) which asserts that π_1 is a closed function.

6.18.　(b) Two arguments will be given to establish that any H-slice is closed when S is compact: Since $H_a = L_a \cap R_a$, this result follows directly from part (a) and its analog for R_a.

The second argument goes as follows and it does not use part (a): $\mathscr{H}$ is closed by (6.7) since S is compact. Note that the natural function $n: S \to S/\mathscr{H}$ is continuous since $S/\mathscr{H}$ has the quotient topology. Let $\bar{x}$ be in $S/\mathscr{H}$. Recall, again, that in a Hausdorff space singleton sets are closed and that $S/\mathscr{H}$ is Hausdorff by (5.6) because S is compact and $\mathscr{H}$ is closed. Therefore, $n^{-1}(\bar{x}) = H_x$ is a closed set in S by (1.1).

6.19   By (6.3) $\mathcal{L}_o$ is closed. S being compact implies S × S is compact by the Tychonoff Product Theorem and so $\mathcal{L}_o$ is compact by (1.7). Since (6.4) states that τ is continuous, then $\tau\,\mathcal{L}_o$ is compact by (1.11); and, since S × S is Hausdorff by (1.5), then $\tau\mathcal{L}_o$ is also closed using (1.8). Therefore, $\mathcal{L}_o \cap \tau\,\mathcal{L}_o = \mathcal{L}$ is closed since the intersection of two closed sets is closed by (i) of the basic definition of a topology and one of DeMorgan's Laws.

I N D E X